WOODY PLANTS OF THE NORTHERN FOREST

A PHOTOGRAPHIC GUIDE

JERRY JENKINS

First published 2018 by Cornell University Press

Printed in China

Library of Congress Cataloging-in-Publication Data
Names: Jenkins, Jerry (Jerry C.), author.
Title: Woody plants of the northern forest : a photographic guide / Jerry Jenkins.
Description: Ithaca : Comstock Publishing Associates, a division of Cornell University Press, 2018. | Series: A northern forest atlas guide | Includes index.
Identifiers: LCCN 2017036242 | ISBN 9781501719684 (pbk. ; alk. paper)
Subjects: LCSH: Woody plants—Northeastern States—Identification. | Woody plants—Canada, Eastern—Identification. | Forest plants—Northeastern States—Identification. | Forest plants—Canada, Eastern—Identification.
Classification: LCC QK118 .J46 2018 | DDC 582.16—dc23
LC record available at https://lccn.loc.gov/2017036242

Cornell University Press strives to use environmentally responsible suppliers and materials to the fullest extent possible in the publishing of its books. Such materials include vegetable-based, low-VOC inks and acid-free papers that are recycled, totally chlorine-free, or partly composed of nonwood fibers. For further information, visit our website at cornellpress.cornell.edu.

CONTENTS

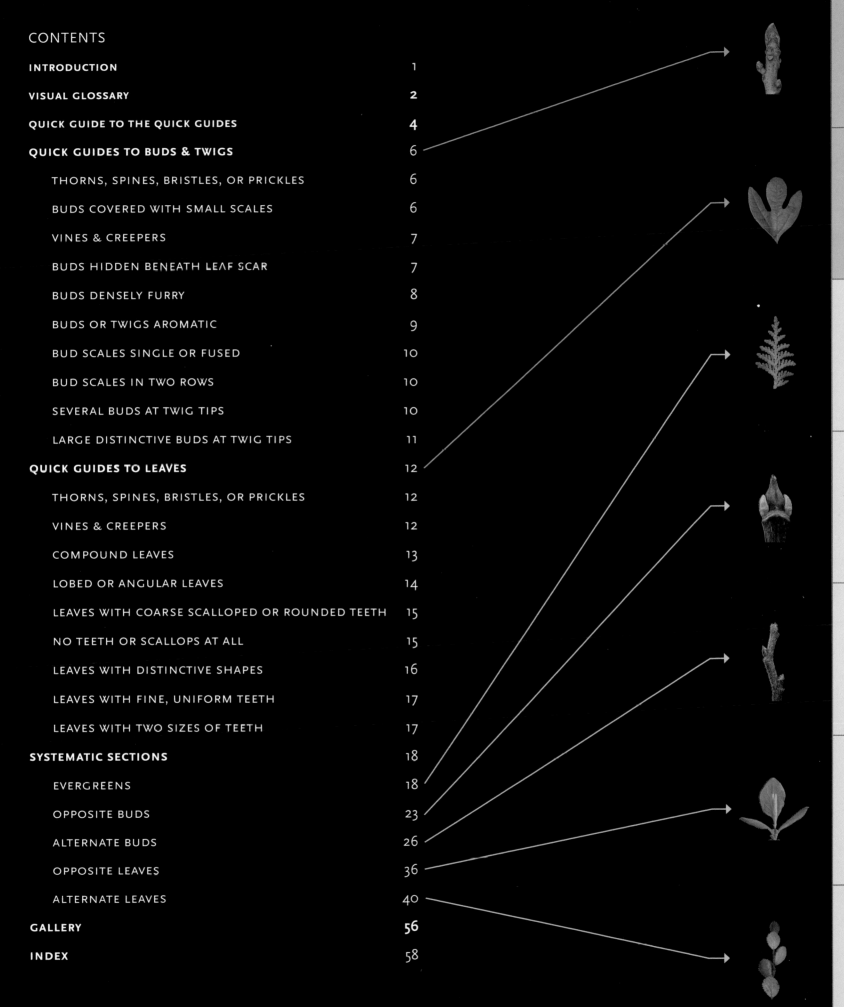

QUICK GUIDES TO BUDS & TWIGS

QUICK GUIDES TO LEAVES

EVERGREENS

OPPOSITE BUDS

ALTERNATE BUDS

OPPOSITE LEAVES

ALTERNATE LEAVES

The Northern Forest Region, showing some Atlas study sites

THE NORTHERN FOREST REGION lies between the oak forests of the eastern United States and the boreal forests of eastern Canada. It is, loosely, the region where forests are dominated by mixtures of maple, birch, beech, spruce, hemlock, and pine. It is also, collectively, one of the largest and most continuous temperate forests left in the world.

The northern forest and its associated communities contain about 265 species of woody plants. We illustrate about 235 of them here. The remaining species, mostly western, have not been photographed yet.

This guide is intended for the rapid identification of twigs and leaves. It will suggest and eliminate but not confirm. For easy plants it may be all you need. For hard ones, it is a good start.

Structurally, it contains nineteen quick guides and five systematic sections. The quick guides illustrate the species that share a distinctive feature—thorns, say, or lobed leaves. Not every species has distinctive features, so not every species is in a quick guide. The systematic sections present the species in each of five basic groups (evergreens, opposite buds, alternate buds, opposite leaves, and alternate leaves), which are arranged alphabetically by family and genus.

The treatments here are intentionally brief. Each species is reduced to two or three images and a few annotations. For more images, see our collection online at northernforestatlas.org, or download our *Digital Atlas of the Woody Plants of the Northern Forest,* which has high-resolution images of bark, buds, flowers, fruits, and leaves. For natural history and ecology, see Welby R. Smith's fine *Trees and Shrubs of Minnesota,* or wait a year or two for our field guide to woody plants.

The text and annotations are largely self-explanatory. A visual glossary of common descriptive terms appears on pages 2 and 3, and a guide to the quick guides on pages 4 and 5. Symbols used throughout the book are shown below. Note that paired and whorled leaves and buds are flagged only in quick guides and in the evergreens, where opposite

and alternate arrangements occur together. They not flagged when the whole group is opposite.

Several notes to start you off: First, not every woody plant can be identified by leaves or twigs alone. In the difficult groups you need some combination of twigs, leaves, and bark. In the most difficult groups, you need both leaves and flowers.

Second, many important details are small. I routinely use 7× and 20× lenses in the field and a 6× to 50× stereo microscope at home. They are fun to look through and prevent embarrassing mistakes.

Third, plants vary. Some of the variation is meaningful and some accidental. Either way, the leaf or bud that you are looking at will differ from the photograph in this book, and the next one you look at will differ from that. Your job is to figure out where, for you, casual variation ends and species lines begin. This is fun, but not easy.

Fourth, about the photography. The studio photos in this book are multi-image composites, shot with Canon lenses and a Stackshot rail and stacked with Zerene Stacker. This allows great clarity and depth and is rapidly making single-image macrophotography obsolete. For more details, see the articles on stacking at northernforestatlas.org.

Last, a personal note. The northern forest plants have been my companions for fifty years. They have been good friends and generous teachers. Much of what I know about ecology, beauty, and pattern has come from them. Spend time with them and think about what you see, and they will teach you too.

If you do, please consider how you can help them survive. Every year the world becomes less stable and the human footprint heavier. The northern forests, like much of the biosphere, are at risk. They need stewards and protectors. If this book can convince you to become one, it will have served its purpose and repaid its author's debt.

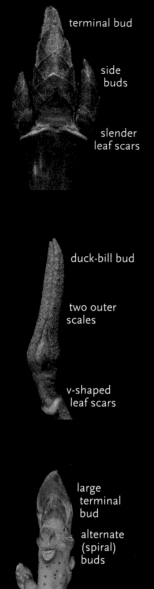

terminal bud

side buds

slender leaf scars

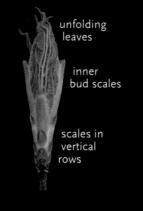

unfolding leaves

inner bud scales

scales in vertical rows

no bud scales

dense fur

stalked buds

opposite (paired) buds

buds on shelves

duck-bill bud

two outer scales

v-shaped leaf scars

bud scales fused

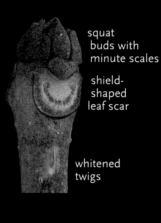

squat buds with minute scales

shield-shaped leaf scar

whitened twigs

buds pressed against twig

base of leafstalk persists

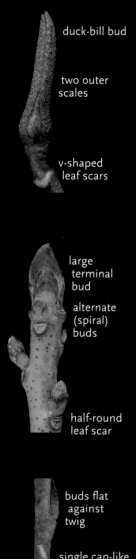

large terminal bud

alternate (spiral) buds

half-round leaf scar

multiple buds

lobed leaf scar

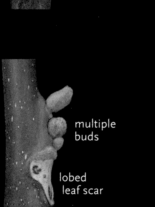

clustered buds

whitened twigs

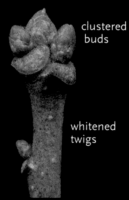

elongate buds

no bud scales

tiny yellow scales

buds flat against twig

single cap-like scale

slender leaf scar

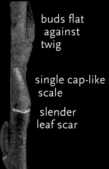

blunt, oval bud

white resin

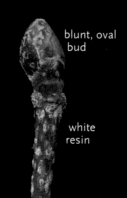

conical bud

aborted tip of twig

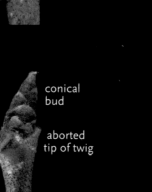

scales in two vertical rows

oval leaf scar

evergreen
broad leaves

edges
rolled under

edges
flat

leaves single
(alternate)

leaves paired
(opposite)

evergreen
needle
leaves

sharp
tips

deciduous
needle leaves

needles
whitened

blunt
tips

long, slender
evergreen
needles

needle
tips

short, flat,
needles,
whitened
above

needles in a
bundle

needles
in threes

long, flat
needles with
rounded tips

needle
bases
round

flattened,
fern-like
branches

paired
scale-
leaves

glossy,
thick,
evergreen

no
teeth

acuminate
tip

curved
veins

short tip

sharp, even
teeth

straight
veins

long tip

two sizes of
teeth

oblong

oval

rounded-oblong

rounded-triangular

rounded
lobes

sharp
notches

lobes with
bristle tips

deep
bays

leaflets
connected
by wing

blunt
tips

oval
leaflets

sharp
teeth

palmately lobed

pinnately (feather) lobed

pinnately compound

palmately compound

3

QUICK GUIDE TO THE QUICK GUIDES

THORNS, SPINES, BRISTLES, OR PRICKLES, p. 6

 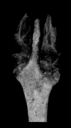 

BUDS OR TWIGS AROMATIC, p. 9

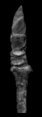  

BUDS COVERED WITH SMALL SCALES, p. 6

BUD SCALES SINGLE OR FUSED, p. 10

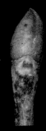  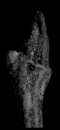

VINES & CREEPERS, p. 7

BUD SCALES IN TWO ROWS, p. 10

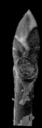 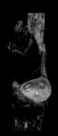 

BUDS HIDDEN BENEATH LEAF SCAR, p. 7

 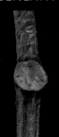 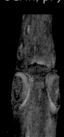

SEVERAL BUDS AT TWIG TIPS, p. 10

BUDS DENSELY FURRY, p. 8

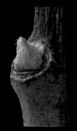  

LARGE, DISTINCTIVE BUDS AT TWIG TIPS, p. 11

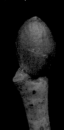 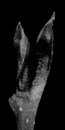 

THORNS, SPINES, BRISTLES, OR PRICKLES, p. 12

NO TEETH OR SCALLOPS AT ALL, p. 15

VINES & CREEPERS, p. 12

LEAVES WITH DISTINCTIVE SHAPES, p. 16

COMPOUND LEAVES, p. 13

LEAVES WITH FINE, UNIFORM TEETH, p. 17

   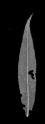

LOBED OR ANGULAR LEAVES, p. 14

    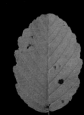

LEAVES WITH TWO SIZES OF TEETH, p. 17

LEAVES WITH COARSE SCALLOPED OR ROUNDED TEETH, p. 15

Buds and leaves from the 19 quick identification groups on pages 6 to 17. Each group has a distinctive feature—hidden buds, lobed leaves, etc. Thorny and viny plants have two groups each, one for twigs and one for leaves. The buds and leaves shown here are samples; many of the groups are more diverse than shown.

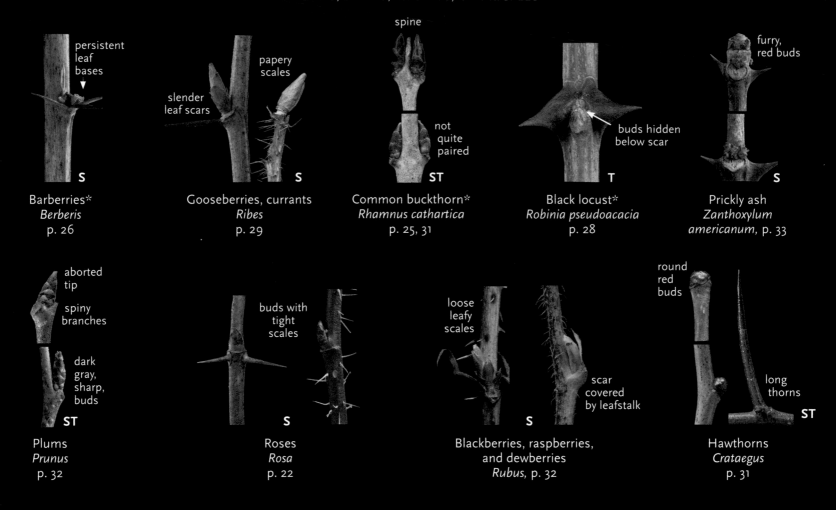

Barberries*
Berberis
p. 26
— persistent leaf bases

Gooseberries, currants
Ribes
p. 29
— slender leaf scars — papery scales

Common buckthorn*
Rhamnus cathartica
p. 25, 31
— spine — not quite paired

Black locust*
Robinia pseudoacacia
p. 28
— buds hidden below scar

Prickly ash
Zanthoxylum americanum, p. 33
— furry, red buds

Plums
Prunus
p. 32
— aborted tip — spiny branches — dark gray, sharp, buds

Roses
Rosa
p. 22
— buds with tight scales

Blackberries, raspberries, and dewberries
Rubus, p. 32
— loose leafy scales — scar covered by leafstalk

Hawthorns
Crataegus
p. 31
— round red buds — long thorns

BUDS COVERED WITH SMALL SCALES

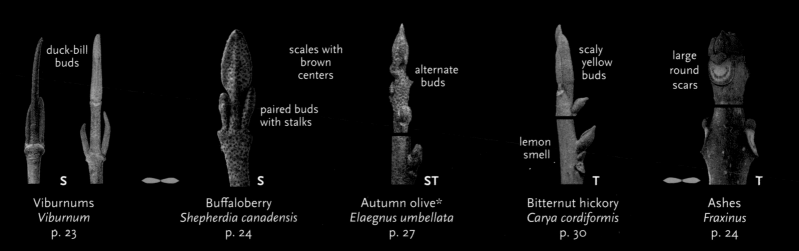

Viburnums
Viburnum
p. 23
— duck-bill buds

Buffaloberry
Shepherdia canadensis
p. 24
— scales with brown centers — paired buds with stalks

Autumn olive*
Elaegnus umbellata
p. 27
— alternate buds

Bitternut hickory
Carya cordiformis
p. 30
— scaly yellow buds — lemon smell

Ashes
Fraxinus
p. 24
— large round scars

The FIRST GROUP includes plants with thorns (modified branch tips) and various sorts of prickles and bristles that arise from the sides of the branches. Some, like those of locust, are associated with leaf scars; others, like those in blackberries, occur all along the branch. Note that the spiny tips of plums may or may not be conspicuous; that barberry buds are short and stubby; visy gooseberries, with similar

prickles, have long buds with papery scales; and that roses have tight scales and clear scars while blackberries and raspberries have loose scales which obscure the scars.

The SECOND GROUP has buds covered with scales. The scales are conspicuous in the buffaloberry and autumn olive. They are tiny and appear like small bumps in the viburnums, bitternut, and the ashes

VINES & CREEPERS

POISONOUS TO TOUCH

climbs with small roots

elongate, furry buds

viny or shrubby

creeps or climbs

Poison ivy
Toxicodendron species
p. 26

hairy twigs

twining

Hairy honeysuckle
Lonicera hirsuta
p. 24

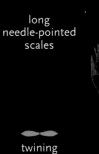

long needle-pointed scales

smooth twigs

papery collar

Limber honeysuckle
Lonicera dioica
p. 24

blunt scales

twining

American bittersweet
Celastrus scandens
p. 27

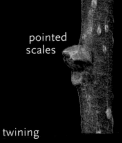

pointed scales

Oriental bittersweet*
Celastrus orbiculatus
p. 27

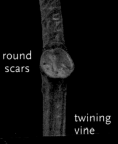

round scars

twining vine

Moonseed
Menispermum canadense, p. 30

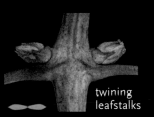

twining leafstalks

Clematis
Clematis
p. 25

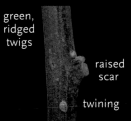

green, ridged twigs

raised scar

twining

Bittersweet nightshade*
Solanum dulcamara
p. 35

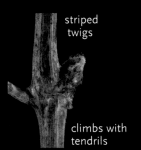

striped twigs

climbs with tendrils

Grapes
Vitis
p. 35

big round scar

climbs or creeps

Virginia creeper
Parthenocissus
p. 35

BUDS HIDDEN BENEATH LEAF SCAR

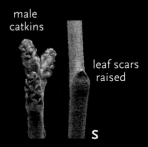

male catkins

leaf scars raised

S

Fragrant sumac
Rhus aromatica
p. 26

may have spines

T

Black locust*
Robinia pseudoacacia
p. 28

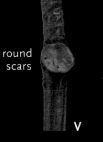

round scars

V

Moonseed
Menispermum canadense
p. 30

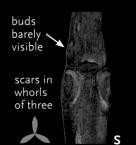

buds barely visible

scars in whorls of three

S

Buttonbush
Cephalanthus occidentalis
p. 25

VINES commonly twine or climb by tendrils or rootlets. A few creep, though they are usually erect at the tips. A few, as an option, may grow as free-standing shrubs. Their ways of climbing differ. The bittersweets, moonseed, and the nightshade climb by twining. The honeysuckles may twine loosely or be free-standing. Clematis climbs by twining leafstalks. The grapes climb by tendrils. Virginia creeper has both tendrils and small rootlets. Poison ivy, depending on the species and the habitat, may creep, climb by rootlets, or grow as a free-standing shrub.

The HIDDEN-BUD SPECIES, have buds are under leaf scars, persistent leaf bases, or bark. All are quite distinctive. They are unrelated and, except for the hidden buds, have no other common features.

QUICK GUIDES TO BUDS & TWIGS

QUICK GUIDES TO LEAVES

EVERGREENS

OPPOSITE BUDS

ALTERNATE BUDS

OPPOSITE LEAVES

ALTERNATE LEAVES

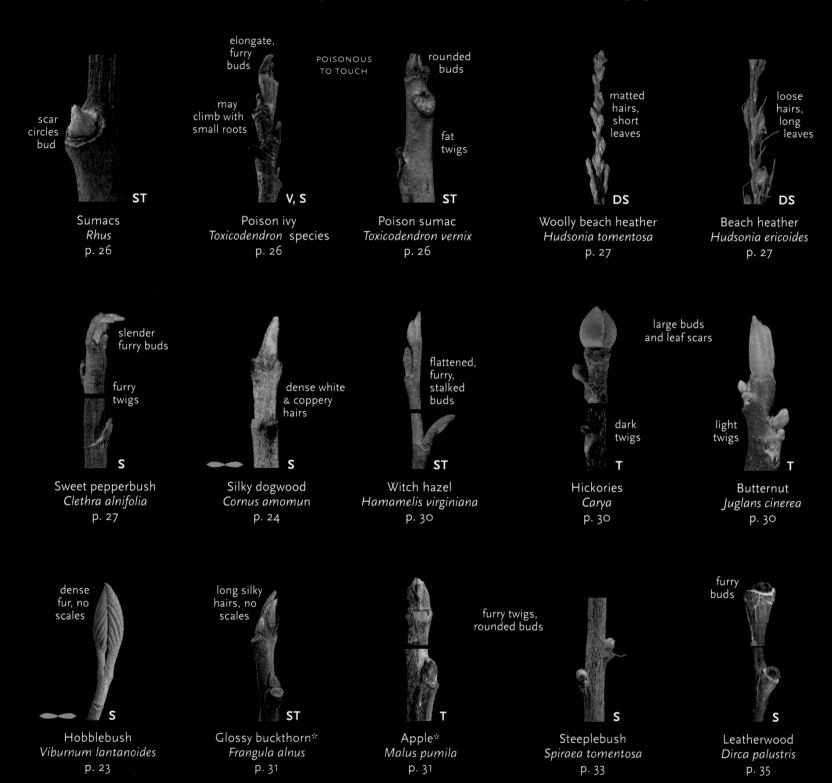

scar circles bud

ST

Sumacs
Rhus
p. 26

elongate, furry buds

may climb with small roots

POISONOUS TO TOUCH

V, S

Poison ivy
Toxicodendron species
p. 26

rounded buds

fat twigs

ST

Poison sumac
Toxicodendron vernix
p. 26

matted hairs, short leaves

DS

Woolly beach heather
Hudsonia tomentosa
p. 27

loose hairs, long leaves

DS

Beach heather
Hudsonia ericoides
p. 27

slender furry buds

furry twigs

S

Sweet pepperbush
Clethra alnifolia
p. 27

dense white & coppery hairs

S

Silky dogwood
Cornus amomun
p. 24

flattened, furry, stalked buds

ST

Witch hazel
Hamamelis virginiana
p. 30

large buds and leaf scars

dark twigs

T

Hickories
Carya
p. 30

light twigs

T

Butternut
Juglans cinerea
p. 30

dense fur, no scales

S

Hobblebush
Viburnum lantanoides
p. 23

long silky hairs, no scales

ST

Glossy buckthorn✳
Frangula alnus
p. 31

furry twigs, rounded buds

T

Apple✳
Malus pumila
p. 31

S

Steeplebush
Spiraea tomentosa
p. 33

furry buds

S

Leatherwood
Dirca palustris
p. 35

FURRY BUDS include some plants (hobblebush, sweet pepperbush, witch hazel, and glossy buckthorn) that lack scales altogether and others (steeplebush, butternut, ...) in which the scales are obscured by dense fur and not visible. Leatherwood and the sumacs have leaf scars that surround the buds. Poison ivy and poison sumac, both potent skin irritants, can be recognized by the combination of shield-shaped leaf scars and small furry buds. Poison ivy is a vine or low shrub with slender twigs. Poison sumac is a tall shrub with fatter ones. The hickories have dark outer scales without fur and light-tan inner ones with a dense covering of tiny, multi-branched hairs. In some species the dark scales persist through the winter. In others they fall early.

——————————— BUDS OR TWIGS AROMATIC ———————————

QUICK GUIDES TO
BUDS & TWIGS

QUICK GUIDES
TO LEAVES

EVERGREENS

OPPOSITE BUDS

ALTERNATE BUDS

OPPOSITE LEAVES

ALTERNATE LEAVES

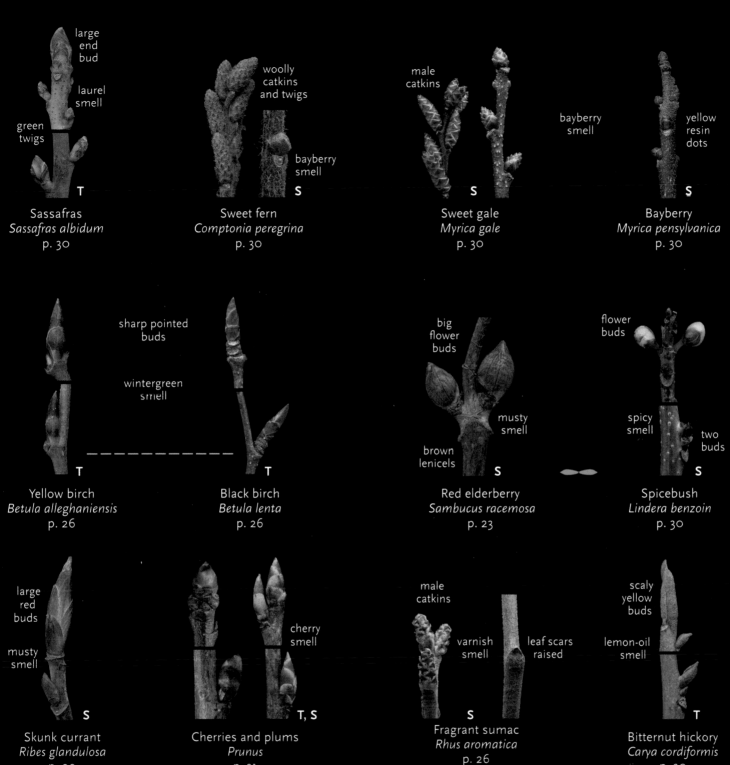

large
end
bud

laurel
smell

green
twigs

T

Sassafras
Sassafras albidum
p. 30

woolly
catkins
and twigs

bayberry
smell

S

Sweet fern
Comptonia peregrina
p. 30

male
catkins

S

Sweet gale
Myrica gale
p. 30

bayberry
smell

yellow
resin
dots

S

Bayberry
Myrica pensylvanica
p. 30

sharp pointed
buds

wintergreen
smell

T

Yellow birch
Betula alleghaniensis
p. 26

T

Black birch
Betula lenta
p. 26

big
flower
buds

musty
smell

brown
lenicels

S

Red elderberry
Sambucus racemosa
p. 23

flower
buds

spicy
smell

two
buds

S

Spicebush
Lindera benzoin
p. 30

large
red
buds

musty
smell

S

Skunk currant
Ribes glandulosa
p. 29

cherry
smell

T, S

Cherries and plums
Prunus
p. 31

male
catkins

varnish
smell

leaf scars
raised

S

Fragrant sumac
Rhus aromatica
p. 26

scaly
yellow
buds

lemon-oil
smell

T

Bitternut hickory
Carya cordiformis
p. 30

AROMATIC TWIGS In most of these species the smell is throughout the twig and released when it is broken or crushed. Make sure that you know poison ivy or poison sumac (p. 8) before you start breaking and smelling. In the poplars the smell comes from the liquid yellow resin in the buds. In bayberry, sweet gale, and sweet fern, it comes from the dots of yellow resin on the leaves and twigs.

The smells themselves are distinctive and easily learned: wintergreen in the birches, sweet-spicy and pungent in sassafras and spicebush, musty-spicy in skunk currant and elderberry, oily-lemon in bitternut, varnish-like in fragrant sumac, resinous and bayberry-like in bayberry, sweet gale, and sweet fern, and cyanic (like cherry pits) in the cherries and plums.

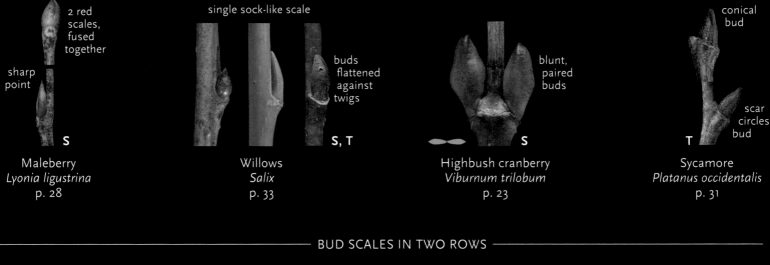

2 red scales, fused together

sharp point

S

Maleberry
Lyonia ligustrina
p. 28

single sock-like scale

buds flattened against twigs

S, T

Willows
Salix
p. 33

blunt, paired buds

S

Highbush cranberry
Viburnum trilobum
p. 23

conical bud

scar circles bud

T

Sycamore
Platanus occidentalis
p. 31

─────── BUD SCALES IN TWO ROWS ───────

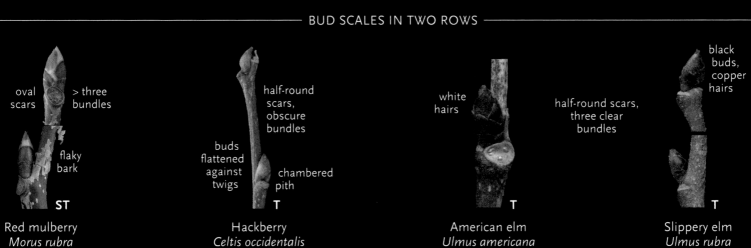

oval scars

> three bundles

flaky bark

ST

Red mulberry
Morus rubra
p. 30

half-round scars, obscure bundles

buds flattened against twigs

chambered pith

T

Hackberry
Celtis occidentalis
p. 30

white hairs

T

American elm
Ulmus americana
p. 35

black buds, copper hairs

half-round scars, three clear bundles

T

Slippery elm
Ulmus rubra
p. 35

─────── SEVERAL BUDS AT TWIG TIPS ───────

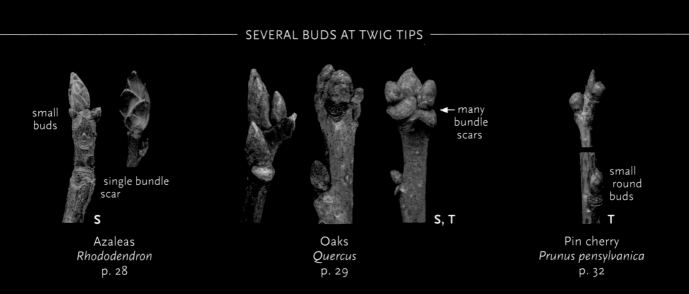

small buds

single bundle scar

S

Azaleas
Rhododendron
p. 28

← many bundle scars

S, T

Oaks
Quercus
p. 29

small round buds

T

Pin cherry
Prunus pensylvanica
p. 32

SINGLE AND FUSED BUD SCALES Willow buds and sycamore are covered by a single scale with its edges more-or-less fused. Maleberry buds are included because they look like a small willow; they actually have two outer scales whose edges can be hard to see. Note that the leaf-scars of willows are slender curved lines while those of maleberry are small ovals or triangles.

BUD SCALES IN TWO ROWS These buds are somewhat flattened. Facing the flat side, they have alternating scales in two rows, one on the left and one on the right, like veins of a feather. Most of our other plants with alternate leaves have bud scales that spiral around the bud. In many of them, the buds are also arranged in two rows on the twigs.

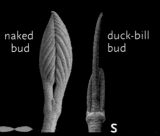

naked
bud

duck-bill
bud

S

Viburnums
Viburnum
p. 23

end
buds
stalked

S

Alders
Alnus
p. 26

small
buds
below
large
one

S

Azaleas
Rhododendron
p. 28

long,
sharp
buds

T

Beech
Fagus grandifolia
p. 29

musty
small

S

Skunk currant
Ribes glandulosa
p. 29

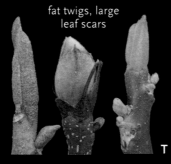

fat twigs, large
leaf scars

T

Hickories and walnuts
Carya & Juglans
p. 30

green
aromatic
twigs

T

Sassafras
Sassafras albidum
p. 30

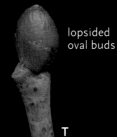

lopsided
oval buds

T

Basswood
Tilia americana
p. 30

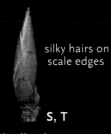

silky hairs on
scale edges

S, T

Shadbushes
Amelanchier
p. 31

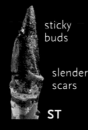

sticky
buds

slender
scars

ST

Mountain ashes
Sorbus
p. 33

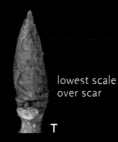

lowest scale
over scar

T

Poplars
Populus
p. 33

duck-bill
bud

ST

Striped maple
Acer pensylvanicum
p. 25

flattened,
furry buds

ST

Witch hazel
Hamamelis virginiana
p. 30

SEVERAL BUDS AT TIPS Oaks typically have club-shaped twig tips and several buds of about the same size. Azaleas are less clubbed and often have a large flower bud with several smaller leaf buds at its base. Both groups vary. The best characters for azaleas are the small triangular leaf scars with one bundle scar, and for oaks the larger oval ones with five or more.

DISTINCTIVE BUDS: A loose group of species in which the end buds are large (often 1 cm or more), and (often) the side buds smaller. Many other species, not included, have large buds in spring.

CK GUIDES TO
DS & TWIGS

QUICK GUIDES
TO LEAVES

EVERGREENS

OPPOSITE BUDS

ALTERNATE BUDS

OPPOSITE LEAVES

ALTERNATE LEAVES

S

Barberries*
Berberis
p. 26

S

Gooseberry
Ribes
p. 45

S

Prickly ash
Zanthoxylum americanum
p. 51

S

Roses
Rosa
p. 48

T

Black locust*
Robinia pseudoacacia
p. 44

ST

Common buckthorn*
Rhamnus cathartica
p. 47

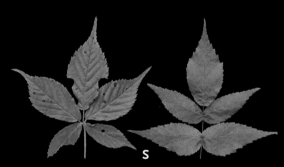

S

Blackberries, raspberries and dewberries
Rubus
p. 48

ST

Hawthorns
Crataegus species
p. 50

Plums
Prunus
p. 50
ST

POISONOUS
TO TOUCH

Poison ivy
Toxicodendron species
p. 40

Honeysuckle
Lonicera
p. 36

Bittersweet
Celastrus
p. 42

Moonseed
*Menispermum
canadense*, p. 47

Clematis
Clematis
p. 38

Bittersweet nightshade*
Solanum dulcamara
p. 54

Grapes
Vitis
p. 54

Virginia creepers
Parthenocissus
p. 55

To facilitate comparison, the quick guides to leaves have small images without annotations or notes. Use the page
references to jump to the Systematic Section for larger images and detailed notes and annotations.

QUICK GUIDES TO
BUDS & TWIGS

QUICK GUIDES
TO LEAVES

EVERGREENS

OPPOSITE BUDS

ALTERNATE BUDS

OPPOSITE LEAVES

ALTERNATE LEAVES

—————————————— COMPOUND LEAVES ——————————————

ST

Sumacs
Rhus p. 40

T

Walnuts
Juglans
p. 46

T

Ashes
Fraxinus
p. 51

S

Prickly ash
*Zanthoxylum
americanum*, p. 51

ST

Mountain ashes
Sorbus
p. 49

S

Elderberries
Sambucus
p. 36

S

Shrubby cinquefoil
Dasiphora fruticosa, p. 49

S

Roses
Rosa
p. 48

S

Leadplant*
Amorpha fruticosa
p. 44

T

Black locust*
Robinia pseudoacacia
p. 44

T

Hickories
Carya
p. 46

POISONOUS
TO TOUCH

ST

Poison sumac
Toxicodendron vernix
p. 40

S,V

Poison ivy
Toxicodendron
p. 40

S

Blackberries, raspberries and dewberries
Rubus
p. 48

T

Box elder
Acer negundo
p. 39

S

Bladdernut
Staphylea trifolia
p. 39

V

Clematis
Clematis
p. 38

S

Virginia creepers
Parthenocissus
p. 55

T   T   S, ST   T   T

Oaks, *Quercus*, p. 44

T

Sassafras
Sassafras albidum
p. 46

S

Gooseberry and Currants
Ribes
p. 45

S

Viburnums
Viburnum
p. 36

V

Moonseed
Menispermum canadense,
p. 47

S

Ninebark
Physocarpus opulifolius
p. 50

T

ST

Maples
Acer
p. 39

T

V

Bittersweet nightshade*
Solanum dulcamara
p. 54

ST

Red mulberry
Morus rubra
p. 47

ST

Hawthorns
Crataegus
p. 50

S

Flowering raspberry
Rubus odoratus
p. 49

V

Grapes
Vitis
p. 54

QUICK GUIDES TO BUDS & TWIGS

QUICK GUIDES TO LEAVES

EVERGREENS

OPPOSITE BUDS

ALTERNATE BUDS

OPPOSITE LEAVES

ALTERNATE LEAVES

──────── LEAVES WITH COARSE SCALLOPED OR ROUNDED TEETH ────────

S

Speckled alder
Alnus incana
p. 41

S, T

Oaks
Quercus
p. 44

ST

Chestnut
Castanea dentata
p. 44

T

Beech
Fagus grandifolia
p. 44

ST

Witch hazel
Hamamelis virginiana
p. 46

S

Sweet fern
Comptonia peregrina
p. 47

──────── NO TEETH OR SCALLOPS AT ALL ────────

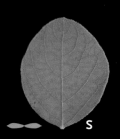

S

Snowberry
Symphoricarpos albus
p. 37

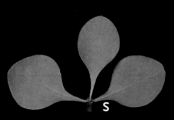

S

Japanese barberry*
Berberis thunbergii
p. 40

T

Black gum
Nyssa sylvatica
p. 47

S, ST

Dogwoods
Cornus
p. 37, 42

S

Leatherwood
Dirca palustris
p. 54

S

Mountain holly
Ilex mucronata
p. 40

S

Buttonbush
*Cephalanthus
occidentalis*, p. 38

S

Blueberries, huckleberries, etc.
Vaccinium, Gaylussacia, . . .
p. 42

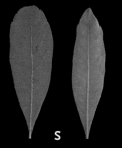

S

Sweet gale, bayberry
Myrica
p. 47

S

Willows
Salix
p. 52

S, T

Spicebush and sassafras
Lindera & Sassafras
p. 46

S, V

Honeysuckles
Lonicera
p. 36

15

S **S** **S** **T**

Arrowwood
Viburnum dentatum
p. 36

Hobblebush
Viburnum lantanoides
p. 36

Round-leaved dogwood
Cornus rugosa
p. 38

Basswood
Tilia americana
p. 46

T **T** **T** **T**

Elms
Ulmus
p. 54

Cottonwood
Populus deltoides
p. 51

Big-toothed aspen
Populus grandidentata
p. 51

Hackberry
Celtis occidentalis
p. 42

ST **C** **S** **S**

Red mulberry
Morus rubra
p. 47

Alpine bearberry
Arctous alpina
p. 43

Dangleberry
Gaylussacia frondosa
p. 43

Sweet pepperbush
Clethra alnifolia
p. 42

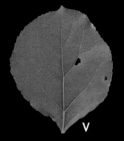

V **T** **S** **T**

Oriental bittersweet*
Celastrus orbiculatus
p. 42

Apple*
Malus pumila
p. 50

Low shadbushes
Amelanchier spicata group
p. 51

Quaking aspen
Populus tremuloides
p. 51

QUICK GUIDES TO
BUDS & TWIGS

QUICK GUIDES
TO LEAVES

EVERGREENS

OPPOSITE BUDS

ALTERNATE BUDS

OPPOSITE LEAVES

ALTERNATE LEAVES

—————————— LEAVES WITH DISTINCTIVE SHAPES ——————————

S

White azalea
*Rhododendron
viscosum*, p. 43

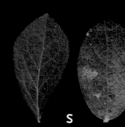

S

Bilberries
Vaccinium
p. 43

C

Willows
Salix
p. 52

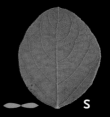

S

Snowberry
Symphoricarpos albus
p. 37

S   **T**   **S**

Willows
Sulix
p. 52

S

Sand cherry
Prunus pumila
p. 50

S

Honeysuckles
Lonicera
p. 36

—————————— LEAVES WITH FINE, UNIFORM TEETH ——————————

T, S

Birches
Betula
p. 41

S

Alders
Alnus
p. 41

ST, S

Shadbushes
Amelanchier
p. 51

T, S

Cherries and Plums
Prunus
p. 50

S

Meadowsweet spiraea
Spiraea alba
p. 51

T, S

Willows
Salix
p. 52

—————————— LEAVES WITH TWO SIZES OF TEETH ——————————

S

Alders
Alnus
p 41

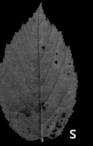

S

Hazelnuts
Corylus
p. 41

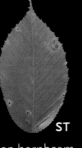

ST

Hop hornbeam
Ostrya virginiana
p. 41

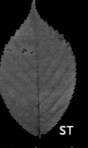

ST

Musclewood
Carpinus caroliniana
p. 41

T

Birches
Betula
p. 41

T

Elms
Ulmus
p. 54

17

CAPRIFOLIACEAE, HONEYSUCKLE FAMILY

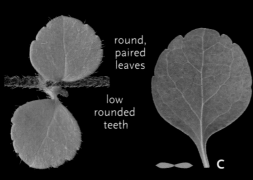

round,
paired
leaves
low
rounded
teeth
C

Twinflower
Linnaea borealis

CISTACEAE, ROCKROSE FAMILY (Dwarf shrubs, on sand)

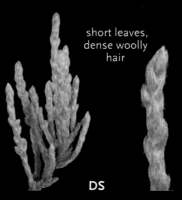

short leaves,
dense woolly
hair
DS

Woolly beach heather
Hudsonia tomentosa

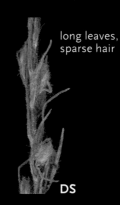

long leaves,
sparse hair
DS

Beach heather
Hudsonia ericoides

CUPRESSACEAE, CYPRESS FAMILY (Scale leaves, in pairs or whorls)

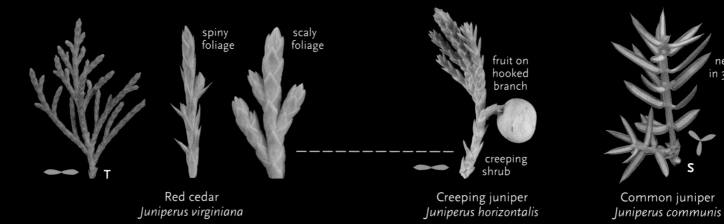

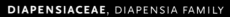

spiny
foliage
scaly
foliage
T

Red cedar
Juniperus virginiana

fruit on
hooked
branch
creeping
shrub

Creeping juniper
Juniperus horizontalis

needles
in 3s
S

Common juniper
Juniperus communis

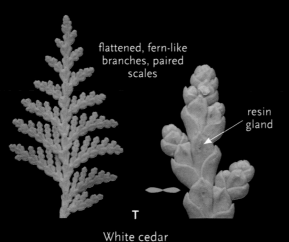

flattened, fern-like
branches, paired
scales
resin
gland
T

White cedar
Thuja occidentalis

DIAPENSIACEAE, DIAPENSIA FAMILY

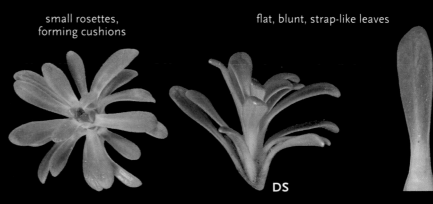

small rosettes,
forming cushions
flat, blunt, strap-like leaves
DS

Diapensia
Diapensia lapponica

EVERGREENS, as used here, are species that have green leaves in winter. (The beach heather is green in early winter.) Most are easily identified. The creeping juniper, a rare northerner, looks like a low version of red cedar; the creeping habit and the curved fruit stalks are the only good characters. The crowberry and the purple mountain heather are low shrubs with fat needles. The crowberry is widely distributed and has untoothed leaves with a narrow gap below. The mountain heather is strictly alpine and has tiny teeth and a wider gap between the inrolled leaf edges on the lower sides of its leaf. The broom crowberry resembles the true crowberry and can grow near it on the north Atlantic coast. It is a lower plant, and best distinguished by the longer, narrow leaves which have a downward pointing bump at their tips.

QUICK GUIDES TO BUDS & TWIGS

QUICK GUIDES TO LEAVES

EVERGREENS

OPPOSITE BUDS

ALTERNATE BUDS

OPPOSITE LEAVES

ALTERNATE LEAVES

ERICACEAE, HEATH FAMILY (Diverse group of shrubs and creepers; common in peatlands, barrens, and tundra) ————————

flattened leaves, needle tips and teeth

DS

Moss heather
Cassiope (Harrimanella) hypnoides

wintergreen smell

thick, waxy

DS

Wintergreen
Gaultheria procumbens

notched tip

black hairs below

DS

Mountain cranberry
Vaccinium vitis-idaea

rounded tip

edges flat

DS, C

Large cranberry
Vaccinium macrocarpon

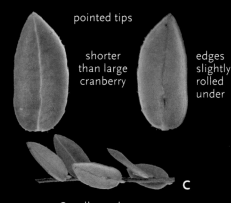

pointed tips

shorter than large cranberry

edges slightly rolled under

C

Small cranberry
Vaccinium oxycoccos

thick, blunt, opposite leaves with inrolled edges

dwarf shrub

Alpine azalea
Loiseluria (Kalmia) procumbens

leaves may appear whorled

minute irregular bumps

thick leaves with inrolled edges

edges meet

DS

Black crowberry
Empetrum nigrum

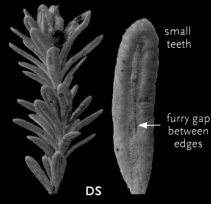

small teeth

furry gap between edges

DS

Purple mountain heather
Phyllodoce caerulea

downward pointing tips

DS

Broom crowberry
Corema conradii

ECOLOGY Most of these species (excepting red cedar, common juniper, and wintergreen) are northern, and many of them are ecological specialists. Twinflower, snowberry, and, in the north, mountain cranberry are boreal forest species. The first two like moist woods and the third, despite its name, relatively dry ones. The broom crowberry are dune species. Bearberry and the junipers are commonest on rocky barrens. Diapensia, crowberry, mountain heather, moss plant, and alpine azalea are species of alpine tundra. Crowberry is common on rocky northern shores and barrens. Leatherleaf and the small cranberry are peatland species. White cedar and the large cranberry are northern generalists; the cedar is most commonly found on base-rich soils, and the cranberry on acid ones.

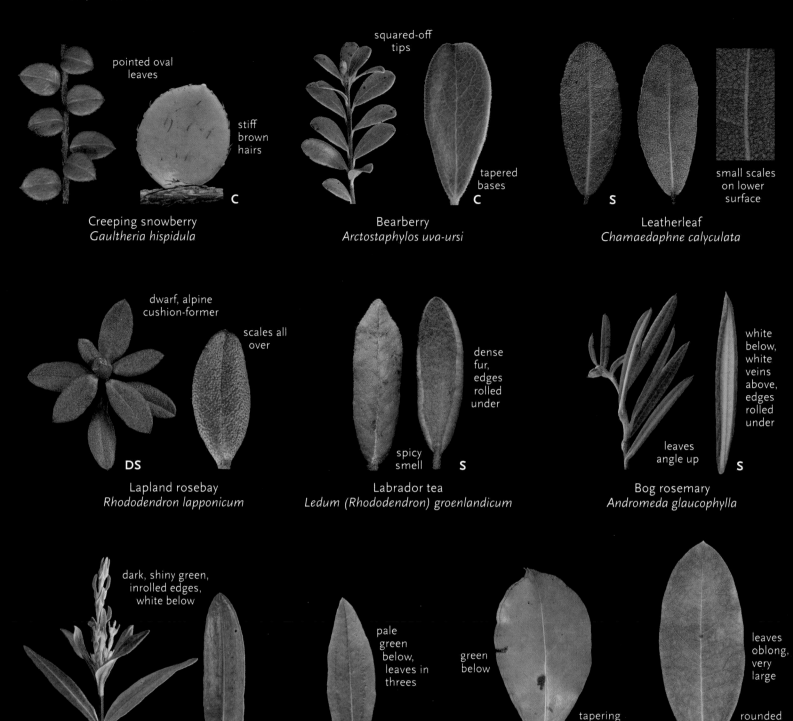

pointed oval
leaves

stiff
brown
hairs

C

Creeping snowberry
Gaultheria hispidula

squared-off
tips

tapered
bases

C

Bearberry
Arctostaphylos uva-ursi

S

small scales
on lower
surface

Leatherleaf
Chamaedaphne calyculata

dwarf, alpine
cushion-former

scales all
over

DS

Lapland rosebay
Rhododendron lapponicum

dense
fur,
edges
rolled
under

spicy
smell

S

Labrador tea
Ledum (Rhododendron) groenlandicum

white
below,
white
veins
above,
edges
rolled
under

leaves
angle up

S

Bog rosemary
Andromeda glaucophylla

dark, shiny green,
inrolled edges,
white below

S

Bog laurel
Kalmia polifolia

pale
green
below,
leaves in
threes

S

Sheep laurel
Kalmia angustifolia

green
below

tapering
to base

S

Mountain laurel
Kalmia latifolia

leaves
oblong,
very
large

rounded
at base

Great laurel
Rhododendron maximum

THE BROAD-LEAVED HEATHS are some of our commonest shrubs. Snowberry, leatherleaf, Labrador tea, bog rosemary, and bog laurel are plants of wet conifer forests and open bogs. Bearberry is found in on open sand or rock, and alpine azalea in alpine tundra. Sheep laurel is a northern generalist found in many sorts of wetlands and forests. Mountain laurel and great laurel are oak-zone species, found mostly to our south but regular near our southern borders. Note that both have alternate leaves, and that the leaves of great laurel are typically larger and more oblong than those of mountain laurel. Pipsissewa, spotted wintergreen, and trailing arbutus (next page) are groundlayer species of oak and conifer forests. All three are most typical of dry sandy soils, but pipsissewa also occurs among mosses in conifer swamps.

ERICACEAE, HEATH FAMILY

big
teeth

thick,
waxy
leaves

DS

Pipsissewa
Chimaphila umbellata

white
stripes

DS

Spotted wintergreen
Chimaphila maculata

oblong leaves,
stiff brown hairs

C

Trailing arbutus
Epigaea repens

PINACEAE, PINE FAMILY (Trees with needle leaves and cones)

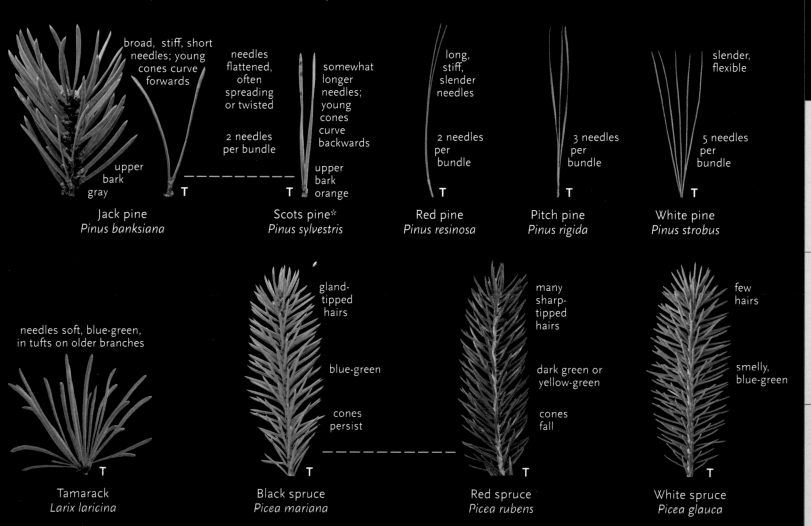

broad, stiff, short
needles; young
cones curve
forwards

upper
bark
gray

T

Jack pine
Pinus banksiana

needles
flattened,
often
spreading
or twisted

2 needles
per bundle

somewhat
longer
needles;
young
cones
curve
backwards

upper
bark
orange

T

Scots pine*
Pinus sylvestris

long,
stiff,
slender
needles

2 needles
per
bundle

T

Red pine
Pinus resinosa

3 needles
per
bundle

T

Pitch pine
Pinus rigida

slender,
flexible

5 needles
per
bundle

T

White pine
Pinus strobus

needles soft, blue-green,
in tufts on older branches

T

Tamarack
Larix laricina

gland-
tipped
hairs

blue-green

cones
persist

T

Black spruce
Picea mariana

many
sharp-
tipped
hairs

dark green or
yellow-green

cones
fall

T

Red spruce
Picea rubens

few
hairs

smelly,
blue-green

T

White spruce
Picea glauca

Our four native pines are separated by their needles. White pine has soft, slender, whitened needles in fives. Pitch pine and red pine have darker and stiffer ones, in threes and twos respectively. Jack and Scots pines have short, flattened, twisted needles in pairs. Those of jack pine run shorter and broader, and it has darker bark and persistent cones.

The pines also differ in geography, though with overlap. Jack pine is the most northern and found on the most infertile soils. Pitch pine is its coastal and Appalachian equivalent. White pine is an Appalachian-Great Lakes species with a wide ecological range; red pine has a similar range, but is a fire-dependent species of ridges and sand plains.

QUICK GUIDES TO
BUDS & TWIGS

QUICK GUIDES
TO LEAVES

EVERGREENS

OPPOSITE BUDS

ALTERNATE BUDS

OPPOSITE LEAVES

ALTERNATE LEAVES

short, flat
needles
with stalks

tiny
teeth

rounded
tips

long, flat needles

needle
tips

strongly
whitened
below

not
strongly
whitened

leaf bases run
down stem

T

Hemlock
Tsuga canadensis

T

Balsam fir
Abies balsamea

S

Canada yew
Taxus canadensis

ROSACEAE, ROSE FAMILY   **SANTALACEAE**, SANDALWOOD FAMILY

compound leaves,
toothed at tip

woody
only at base

DS

Three-toothed cinquefoil
Sibbaldiopsis tridentata

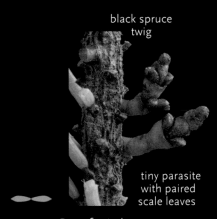

black spruce
twig

tiny parasite
with paired
scale leaves

Dwarf mistletoe
Arceuthobium pusillum

OUR THREE NATIVE SPRUCES, p. 21, are best separated by the hairs on the twigs. Look at the youngest twigs and use a 20x lens. White spruce, which is common only in the northern parts of the northern forest region, is either hairless or has tiny hairs along the grooves between the needle bases. Red spruce, the common large spruce of eastern mountains, and black spruce, the small or medium-sized spruce of peatlands and mountain summits, have hairs all over the needle bases.

BLACK SPRUCE AND RED SPRUCE overlap in many features, particularly cone size and needle length, though black spruce may average a bit shorter on both. The most useful characters for black spruce are the long-persistent cones, the blue-green color of the young needles, and the presence of gland-tipped hairs near the twig tips. Red spruce, in contrast, drops its cones each year, is generally darker green (though often yellowish in wetlands), and has sharp-tipped hairs. The two seem to hybridize at the edges of wetlands.

FIRS, YEWS, AND HEMLOCKS all have flattened needles with white stripes (stomatal bands) on the lower surfaces. Hemlock needles are the shortest of the three and have tiny teeth and stalks. Fir needles have rounded tips and attach by a small round base, like a suction cup. Yew needles have sharper tips and long bases that run down the twigs. Hemlock and fir are trees, yew a sprawling shrub.

CINQUEFOIL AND MISTLETOE are common evergreens, though barely shrubs. Cinquefoil is a low plant of exposed rocky barrens and summits. The leaves are persistent, but only the base of the stem is woody. Mistletoe is a parasite, mostly less than a centimeter high, that grows between the needles of black spruce. It has paired, fleshy leaves that join at their edges and form a cup around the stem. The bumps coming out of the cup are short lateral shoots that bear flowers. The stems retain their scale leaves for several years, making it an evergreen shrub by definition, if not by size.

ADOXACEAE, MOSCHATEL FAMILY (Shrubs with small flowers, toothed or lobed leaves, and 3 or more bundle scars) ————————

dense
fur,
no
scales

S

Hobblebush
Viburnum lantanoides

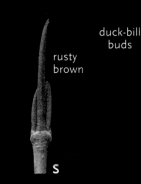

duck-bill
buds

rusty
brown

S

Wild raisin
Viburnum cassinoides

pink-
gray

S

Nannyberry
Viburnum lentago

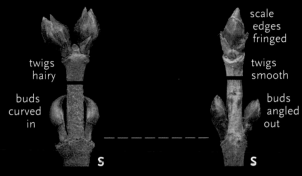

twigs
hairy

buds
curved
in

S

Maple-leaved viburnum
Viburnum acerifolium

scale
edges
fringed

twigs
smooth

buds
angled
out

S

Rafinesque's viburnum
Viburnum rafinesquianum

twigs
ridged,
light tan,
without
hairs

S

Arrowwood
Viburnum dentatum

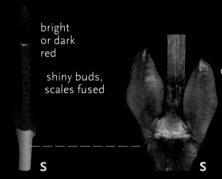

bright
or dark
red

shiny buds,
scales fused

S

Squashberry
Viburnum edule

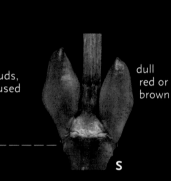

dull
red or
brown

S

Highbush cranberry
Viburnum trilobum

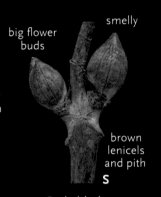

smelly

big flower
buds

brown
lenicels
and pith

S

Red elderberry
Sambucus racemosa

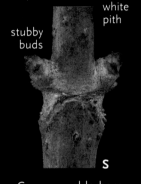

white
pith

stubby
buds

S

Common elderberry
Sambucus canadensis

CAPRIFOLIACEAE, HONEYSUCKLE FAMILY (Shrubs with tubular flowers, 1 bundle scar, and simple, mostly untoothed leaves) ————————

blunt,
fleshy,
mottled
scales

buds point out

S

Canada honeysuckle
Lonicera canadensis

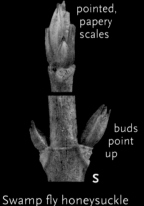

pointed,
papery
scales

buds
point
up

S

Swamp fly honeysuckle
Lonicera oblongifolia

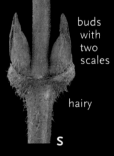

buds
with
two
scales

hairy

S

Mountain fly honeysuckle
Loniceru villosa

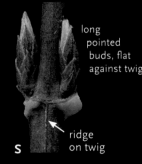

long
pointed
buds, flat
against twig

ridge
on twig

S

Bush honeysuckle
Diervilla lonicera

VIBURNUM TWIGS tend to be tan or gray and to have slender v-shaped or crescentic leaf scars. The buds may be scaleless, have fused scales, a pair of elongate scales, or four or more scales. Three species pairs (nannyberry-wild raisin, maple leaved-Rafinesque's, and squashberry-highbush cranberry) are close, and may not be separable in the winter.

THE HONEYSUCKLES—*Lonicera*—have triangular leaf scars with a single bundle scar that is often obscured by remnants of the leafstalk. The outer scales of their terminal buds often persist around the bases of the twigs. Most are identifiable by their buds, though hairy and limber honeysuckles, p. 24, are quite similar and can cause problems.

QUICK GUIDES TO BUDS & TWIGS

QUICK GUIDES TO LEAVES

EVERGREENS

OPPOSITE BUDS

ALTERNATE BUDS

OPPOSITE LEAVES

ALTERNATE LEAVES

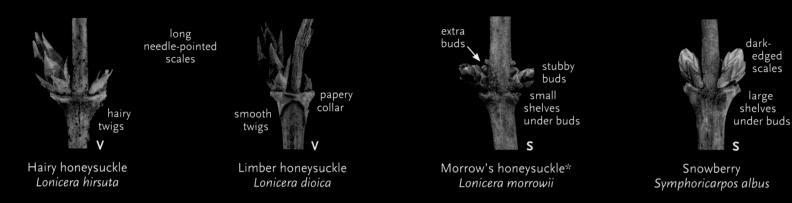

long
needle-pointed
scales

extra
buds

dark-
edged
scales

stubby
buds

hairy
twigs

papery
collar

small
shelves
under buds

large
shelves
under buds

smooth
twigs

V

V

S

S

Hairy honeysuckle
Lonicera hirsuta

Limber honeysuckle
Lonicera dioica

Morrow's honeysuckle*
Lonicera morrowii

Snowberry
Symphoricarpos albus

CORNACEAE, DOGWOOD FAMILY: *CORNUS*, DOGWOODS (Opposite-leaved shrubs and small trees; leaf scars raised on persistent leaf bases)

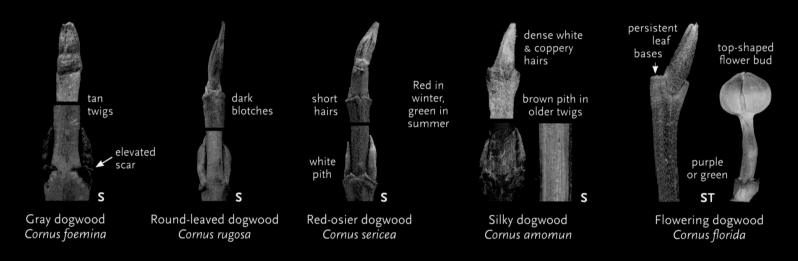

dense white
& coppery
hairs

persistent
leaf
bases

top-shaped
flower bud

tan
twigs

dark
blotches

short
hairs

Red in
winter,
green in
summer

brown pith in
older twigs

elevated
scar

white
pith

purple
or green

S

S

S

S

ST

Gray dogwood
Cornus foemina

Round-leaved dogwood
Cornus rugosa

Red-osier dogwood
Cornus sericea

Silky dogwood
Cornus amomun

Flowering dogwood
Cornus florida

ELAEGNACEAE,
OLEASTER FAMILY

OLEACEAE, OLIVE FAMILY: *FRAXINUS*, ASH (Trees with squat granular buds and large scars)

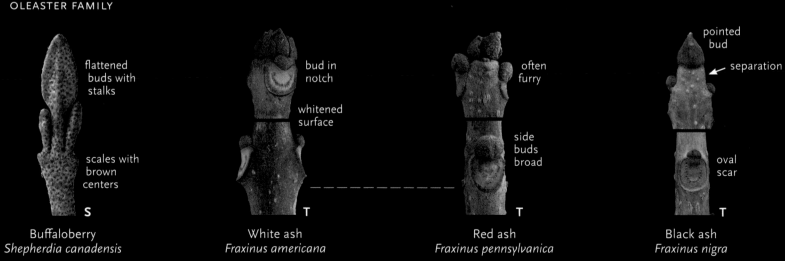

pointed
bud

flattened
buds with
stalks

bud in
notch

often
furry

separation

whitened
surface

side
buds
broad

scales with
brown
centers

oval
scar

S

T

T

T

Buffaloberry
Shepherdia canadensis

White ash
Fraxinus americana

Red ash
Fraxinus pennsylvanica

Black ash
Fraxinus nigra

THE DOGWOODS have slender, somewhat compressed buds and shriveled leaf scars elevated on the persistent bases of the leafstalks. All are to some extent hairy. The best distinctions are the colors of the twigs and the pith from older twigs. Mountain maple, p. 25, much resembles silky dogwood but has clear leaf scars and lacks persistent leafstalk bases.

THE ASHES all have squat, minutely granular or scaly buds and large leaf scars. Red ash and white ash twigs are similar and not always separable. The whitened surface and deeply notched leaf scars are the best marks for white ash, the fur and shallowly notched scars the best marks for red. Neither are completely reliable.

QUICK GUIDES TO BUDS & TWIGS

QUICK GUIDES TO LEAVES

EVERGREENS

OPPOSITE BUDS

ALTERNATE BUDS

OPPOSITE LEAVES

ALTERNATE LEAVES

RANUNCULACEAE, BUTTERCUP FAMILY

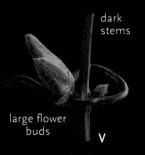

dark stems

large flower buds

V

Purple clematis
Clematis occidentalis

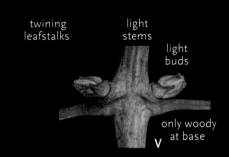

twining leafstalks

light stems

light buds

only woody at base

V

Virgin's bower
Clematis virginiana

RHAMNACEAE, BUCKTHORN FAMILY

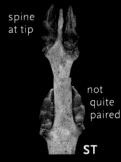

spine at tip

not quite paired

ST

Common buckthorn٭
Rhamnus cathartica

RUBIACEAE, MADDER FAMILY

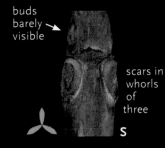

buds barely visible

scars in whorls of three

S

Buttonbush
Cephalanthus occidentalis

SANTALACEAE, SANDALWOOD FAMILY

black spruce twig

tiny parasite with paired scale leaves

Dwarf mistletoe
Arceuthobium pusillum

STAPHYLEACEAE, BLADDERNUT FAMILY

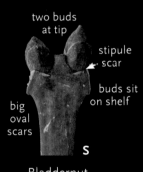

two buds at tip

stipule scar

buds sit on shelf

big oval scars

S

Bladdernut
Staphylea trifolia

SAPINDACEAE, SOAPBERRY FAMILY: *ACER*, MAPLE

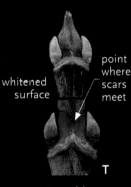

point where scars meet

whitened surface

T

Box elder
Acer negundo

sharp point

warm brown twigs

T

Sugar maple
Acer saccharum

short fat buds

brown twigs

T

Norway maple٭
Acer platanoides

dark red buds and twigs

T

Red and silver maples
Acer rubrum & saccharinum

smooth & shiny

red or green twigs

buds with two visible scales

ST

Striped maple
Acer pensylvanicum

minutely furry

red twigs, short white hairs

ST

Mountain maple
Acer spicatum

THE MAPLES resemble the dogwoods but have more clearly defined scars that are not elevated by leaf-stalk remnants. The buds are often larger and fatter, and the twigs darker or more intensely colored. All except silver and red maple are separable in the winter. Box elder and mountain maple have furry buds, striped maple smooth and shiny ones, sugar maple brown pointed ones, and Norway maple fat rounded ones. Silver and red maples, with short red buds and dark red twigs, are best separated by the drooping branches and the deeply cut leaves of silver maple.

BLADDERNUT looks like an odd shrubby maple with dark green twigs and squat buds that sit on shelves. The terminal bud often aborts, leaving a pair of buds at the tip. The twigs are hairless and the leaf scars large and half round.

25

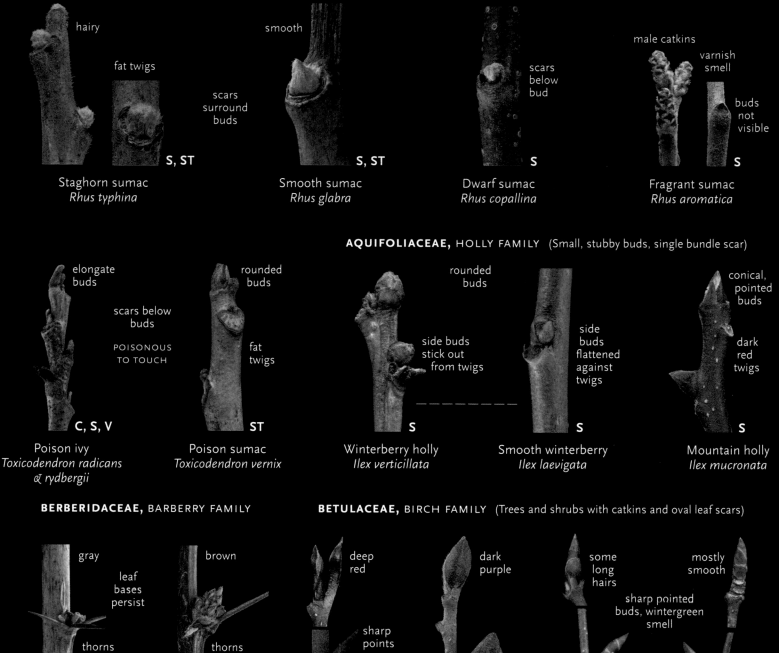

hairy

fat twigs

scars
surround
buds

S, ST

Staghorn sumac
Rhus typhina

smooth

S, ST

Smooth sumac
Rhus glabra

scars
below
bud

S

Dwarf sumac
Rhus copallina

male catkins

varnish
smell

buds
not
visible

S

Fragrant sumac
Rhus aromatica

AQUIFOLIACEAE, HOLLY FAMILY (Small, stubby buds, single bundle scar)

elongate
buds

scars below
buds

POISONOUS
TO TOUCH

C, S, V

Poison ivy
*Toxicodendron radicans
& rydbergii*

rounded
buds

fat
twigs

ST

Poison sumac
Toxicodendron vernix

rounded
buds

side buds
stick out
from twigs

S

Winterberry holly
Ilex verticillata

side
buds
flattened
against
twigs

S

Smooth winterberry
Ilex laevigata

conical,
pointed
buds

dark
red
twigs

S

Mountain holly
Ilex mucronata

BERBERIDACEAE, BARBERRY FAMILY

BETULACEAE, BIRCH FAMILY (Trees and shrubs with catkins and oval leaf scars)

gray

leaf
bases
persist

thorns
often in
threes

S

Common barberry*
Berberis vulgaris

brown

thorns
mostly
single

S

Japanese barberry*
Berberis thunbergii

deep
red

sharp
points

no stalk

S

Green alder
Alnus viridis

dark
purple

stalk

S

Speckled alder
Alnus incana

some
long
hairs

sharp pointed
buds, wintergreen
smell

T

Yellow birch
Betula alleghaniensis

mostly
smooth

T

Black birch
Betula lenta

SUMACS, except for fragrant sumac, have densely hairy buds. In *Rhus* these are more or less surrounded by horseshoe-shaped leaf scars. In *Toxicodendron*, the scars are shield-shaped and below the buds.

WINTERBERRY AND SMOOTH WINTERBERRY ARE close in all their features. If you can find fruits, the ciliate hairs on the sepals of the common winterberry are a good character. The difference in bud shape, shown here, works some of the time.

THE BIRCH FAMILY can often be recognized in winter by the zig-zag twigs, pointed buds, and half-round leaf scars. The shads and cherries, p. 31 and 32, are similar and have to be separated species by species. Within the family, the alders have large buds with a few long scales. Green alder has redder and sharper buds, speckled alder grayer and blunter ones. Additionally, the side buds of speckled alder are stalked.

BETULACEAE, BIRCH FAMILY

  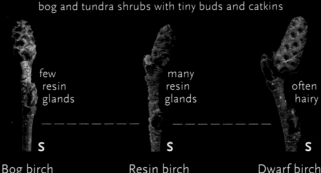  

gummy buds

bumpy twigs

T

Gray birch
Betula populifolia

hairy twigs

T

Paper birch
Betula papyrifera

bog and tundra shrubs with tiny buds and catkins

few resin glands

S

Bog birch
Betula pumila

many resin glands

S

Resin birch
Betula glandulosa

often hairy

S

Dwarf birch
Betula minor

scales with white edges

dark twigs & buds

ST

Musclewood
Carpinus caroliniana

CELASTRACEAE, BITTERSWEET FAMILY

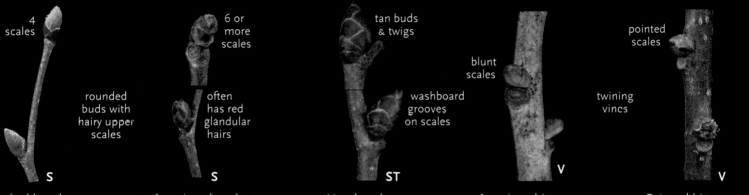

4 scales

rounded buds with hairy upper scales

S

Beaked hazelnut
Corylus cornuta

6 or more scales

often has red glandular hairs

S

American hazelnut
Corylus americana

tan buds & twigs

washboard grooves on scales

ST

Hop hornbeam
Ostrya virginiana

blunt scales

V

American bittersweet
Celastrus scandens

pointed scales

twining vines

V

Oriental bittersweet*
Celastrus orbiculatus

CISTACEAE, ROCKROSE FAMILY   ## CLETHRACEAE, PEPPERBUSH FAMILY   ## CORNACEAE, DOGWOOD FAMILY   ## ELAEGNACEAE, OLEASTER FAMILY

short leaves, dense woolly hair

DS

Woolly beach heather
Hudsonia tomentosa

long leaves, sparse hair

DS

Beach heather
Hudsonia ericoides

slender furry buds

furry twigs

S

Sweet pepperbush
Clethra alnifolia

long, dark shiny scales

leaf bases persist

ST

Alternate-leaved dogwood
Cornus alterniflora

single buds

silver scales with brown centers

ST

Autumn olive*
Elaegnus umbellata

HAZELNUTS have rounded buds with furry upper scales. Beaked hazelnut, the commonest species in the north, typically has four bud scales and stalkless catkins. American hazelnut, taller and more southern, has five or more bud scales, short stalks on the catkins, and coarse red glandular hairs on the twigs. The remaining birches are generally similar. Pick out hop hornbeam by the tan buds with ripples on the scales and musclewood by the dark buds with light-edged scales.

Black and yellow birch, both with a wintergreen smell, are best separated by bark. Gray birch and paper birch are often separably by the glandular bumps on the twigs of gray birch; when in doubt its single catkins and nonpeeling bark are good characters. The three shrubby birches, with short, stubby buds and catkins, have virtually identical twigs and can't be separated in winter.

QUICK GUIDES TO BUDS & TWIGS

QUICK GUIDES TO LEAVES

EVERGREENS

OPPOSITE BUDS

ALTERNATE BUDS

OPPOSITE LEAVES

ALTERNATE LEAVES

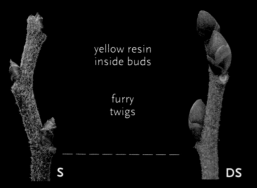

yellow resin
inside buds

furry
twigs

S

Black huckleberry
Gaylussacia baccata

DS

Dwarf huckleberry
Gaylussacia bigeloviana

whitened
twigs

S

Dangleberry
Gaylussacia frondosa

red
scales,
fused
together

sharp
point

S

Maleberry
Lyonia ligustrina

small
buds
below
large

furry
twigs
& buds

S

June pink
Rhododendron prionophyllum

whitened
scales

smooth
twigs

S

Rhodora
Rhododendron canadense

scales with
distinct
points

white
fringes

few long
stiff hairs

S

White azalea
Rhododendron viscosum

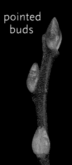

pointed
buds

S

Dwarf bilberry
Vaccinium cespitosum

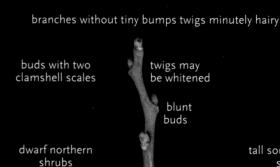

branches without tiny bumps twigs minutely hairy

buds with two
clamshell scales

twigs may
be whitened

blunt
buds

dwarf northern
shrubs

Mountain bilberry
Vaccinium uliginosum

squat,
rounded
buds

tall southern
shrub

Deerberry
Vaccinium stamineum

FABACEAE, BEAN FAMILY

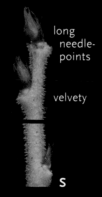

minutely bumpy, red or green twigs, sharp-tipped scales

tall,
often
furry

low,
few
hairs

S

Lowbush blueberries
*Vaccinium angustifolium,
boreale, & pallidum*

S

Highbush blueberry
Vaccinium corymbosum

long
needle-
points

velvety

S

Velvet-leaf blueberry
Vaccinium myrtilloides

pair of
buds

ridged
twigs

S

Leadplant*
Amorpha fruticosa

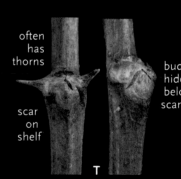

often
has
thorns

scar
on
shelf

buds
hidden
below
scar

T

Black locust*
Robinia pseudoacacia

THE DECIDUOUS HEATHS are hard. As a group, they typically have small triangular scars with a single bundle scar, small leaf buds and larger flower buds. Dangleberry is recognized by the smooth, whitened and somewhat mottled twigs, deerberry by its rounded buds and smooth twigs, maleberry by the slender, sharp-pointed buds, and the azaleas by having several buds at the twig tips. The lowbush blueberries, marked as a group by bumpy, largely hairless twigs, are indistinguishable in winter, and only separable from highbush blueberry by size. The two bilberries may be recognized by their alpine or river-shore habitat, smooth twigs, and small rounded buds. Mountain bilberry is a bigger plant and tends to have three or more scales on some buds. The velvet-leaf blueberry, a boreal species, stands out by its velvety twigs and long-pointed scales. The two huckleberries are very similar, and probably not distinguishable in winter.

FAGACEAE, BEECH FAMILY (Trees and shrubs with oval leaf scars and five or more bundle scars) ─────────────

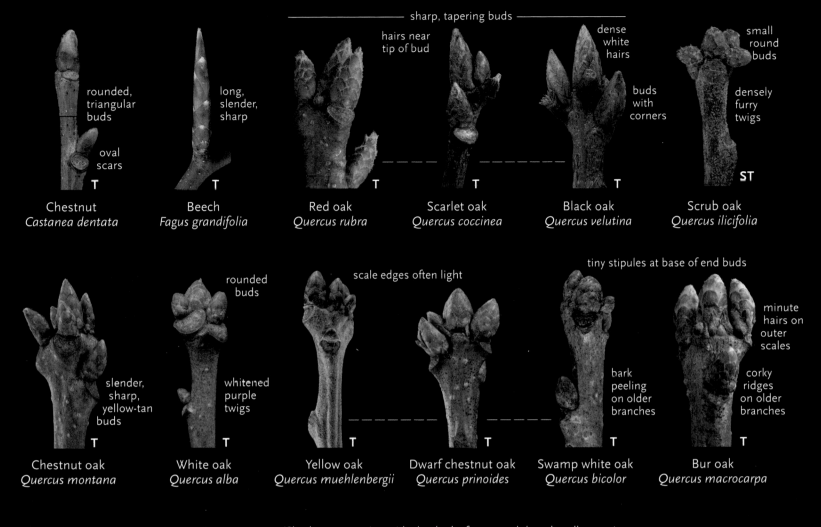

─────────── sharp, tapering buds ───────────

Chestnut
Castanea dentata
— rounded, triangular buds; oval scars — T

Beech
Fagus grandifolia
— long, slender, sharp — T

Red oak
Quercus rubra
— hairs near tip of bud — T

Scarlet oak
Quercus coccinea
— T

Black oak
Quercus velutina
— dense white hairs; buds with corners — T

Scrub oak
Quercus ilicifolia
— small round buds; densely furry twigs — ST

Chestnut oak
Quercus montana
— slender, sharp, yellow-tan buds — T

White oak
Quercus alba
— rounded buds; whitened purple twigs — T

Yellow oak
Quercus muehlenbergii
— scale edges often light — T

Dwarf chestnut oak
Quercus prinoides
— T

Swamp white oak
Quercus bicolor
— tiny stipules at base of end buds; bark peeling on older branches — T

Bur oak
Quercus macrocarpa
— minute hairs on outer scales; corky ridges on older branches — T

GROSSULARIACEAE, GOOSEBERRY FAMILY (Shrubs, some spiny, with slender leaf scars and three bundle scars) ─────────────

Gooseberries
Ribes cynosbati, hirtellum & rotundifolium
— papery scales; spines — S

Swamp black currant
Ribes lacustre
— spiny yellow twigs — S

American currant
Ribes americana
— yellow resin dots; blunt buds — S

Skunk currant
Ribes glandulosa
— large red buds; musty smell — S

Swamp red currant
Ribes triste
— dark buds; low creeping shrub — S

OAKS are best identified by leaves and acorns. Buds and bark are useful, but not sufficient by themselves. With practice you can recognize many of the species from their twigs, as much by their general look as by specific characters: there are, so far as I can see, no clear characters that separate the buds of red and scarlet oaks, or white oak and swamp white oak, or dwarf chestnut and yellow oak.

GOOSEBERRIES (*Ribes cynosbati, hirtellum,* and *rotundifolium*) are closely related, overlap in range and ecology, and can't be reliably identified in winter. Their flowers are fairly distinct. Their leaves vary, in an interesting but confusing way: until someone proves otherwise flowers are required for accurate identification.

QUICK GUIDES TO BUDS & TWIGS
QUICK GUIDES TO LEAVES
EVERGREENS
OPPOSITE BUDS
ALTERNATE BUDS
OPPOSITE LEAVES
ALTERNATE LEAVES

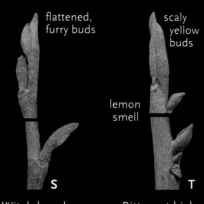

flattened, furry buds

Witch hazel
Hamamelis virginiana

S

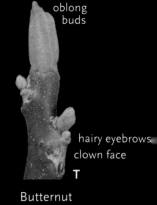

scaly yellow buds

lemon smell

Bitternut hickory
Carya cordiformis

T

large buds

outer scales persist

Shagbark hickory
Carya ovata

T

smaller buds

outer scales fall

Pignut hickory
Carya glabra

T

broad oval

Mockernut hickory
Carya tomentosa

T

oblong buds

hairy eyebrows clown face

Butternut
Juglans cinerea

T

LAURACEAE, LAUREL FAMILY

flower buds

spicy smell

two buds

Spicebush
Lindera benzoin

S

large end bud

green twigs

Sassafras
Sassafras albidum

T

MALVACEAE, MALLOW FAMILY

fat, red or green buds, 2-3 scales

Basswood
Tilia americana

T

MENISPERMACEAE, MOONSEED FAMILY

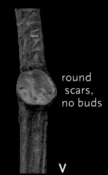

round scars, no buds

Moonseed
Menispermum canadense

V

MORACEAE, MULBERRY FAMILY

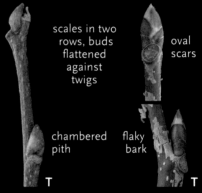

scales in two rows, buds flattened against twigs

chambered pith

Hackberry
Celtis occidentalis

T

oval scars

flaky bark

Red mulberry
Morus rubra

T

MYRICACEAE, BAYBERRY FAMILY (Aromatic shrubs with yellow resin; some with male catkins)

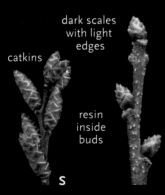

catkins

dark scales with light edges

resin inside buds

Sweet gale
Myrica gale

S

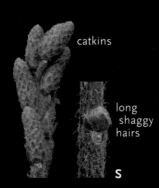

catkins

long shaggy hairs

Sweet fern
Comptonia peregrina

S

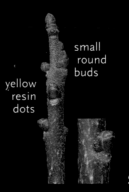

small round buds

yellow resin dots

Bayberry
Myrica pensylvanica

S

NYSSACEAE, TUPELO FAMILY

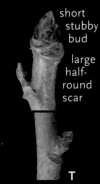

short stubby bud

large half-round scar

Black gum
Nyssa sylvatica

T

SHAGBARK AND PIGNUT HICKORIES are closely related and seem to produce intermediates. Shagbark is best recognized by the loose bark and small hairs near the tips of the leaf teeth. Its buds run larger than those of pignut and tend to keep their dark outer scales longer; neither character is completely reliable. Mockernut is similar to pignut, but

HACKBERRY AND RED MULBERRY resemble elms in their zigzag twigs, oval leaf scars, and bud scales in two vertical rows. Unlike elms, they tend to have relatively light-colored buds and leaf scars with more than three bundle scars. Red mulberry has clear leaf scars, hairless, oval buds and solid pith. Hackberry has small obscure leaf scars, small,

QUICK GUIDES TO BUDS & TWIGS

QUICK GUIDES TO LEAVES

EVERGREENS

OPPOSITE BUDS

ALTERNATE BUDS

OPPOSITE LEAVES

ALTERNATE LEAVES

PINACEAE, PINE FAMILY

PLATANACEAE, PLANE-TREE FAMILY

RHAMNACEAE, BUCKTHORN FAMILY

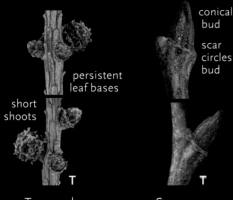

persistent leaf bases

short shoots

T

Tamarack
Larix laricina

conical bud

scar circles bud

T

Sycamore
Platanus occidentalis

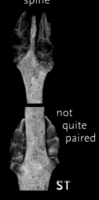

dark scales

buds close to twigs

S

Alder-leaved buckthorn
Rhamnus alnifolia

spine

not quite paired

ST

Common buckthorn*
Rhamnus cathartica

furry buds, no scales

ST

Glossy buckthorn*
Frangula alnus

dark slender outer scales, raised scars

S

New Jersey tea
Ceanothus americanus

ROSACEAE, ROSE FAMILY (Slender or oval leaf scars with three bundle scars)

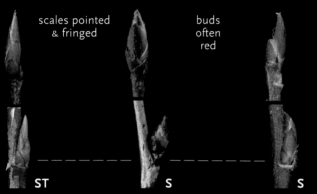

scales pointed & fringed

buds often red

ST

Large shadbushes
*Amelanchier arborea,
canadensis, laevis*

S

Low shadbushes
*Amelanchier
spicata group*

S

Bartram's shadbush
Amelanchier bartramiana

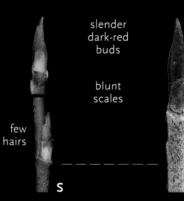

slender dark-red buds

blunt scales

few hairs

S

Black chokeberry
Aronia melanocarpa

matted hairs

S

Red chokeberry
Aronia arbutifolia

big thorns

round red buds

ST

Hawthorns
Crataegus

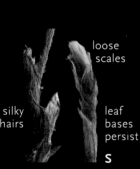

loose scales

silky hairs

leaf bases persist

S

Shrubby cinquefoil
Dasiphora fruticosa

rounded buds

white fur

dark red buds

ST

Apple*
Malus pumila

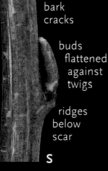

bark cracks

buds flattened against twigs

ridges below scar

S

Ninebark
Physocarpus opulifolius

brown edges to scales

red-brown twigs

T

Black cherry
Prunus serotina

gray edges to scales

gray twigs

S

Choke cherry
Prunus virginiana

THE ROSE FAMILY, pages 31–33, is large and diverse. There are no winter characters for the family as a whole. Several of the genera, particularly the shads, roses, hawthorns, and brambles, hybridize freely and weak species lines. Leaves and flowers are needed to distinguish them to the extent that they can be distinguished at all.

THE SHADS, as a group, can be recognized in the winter by the slender, pointed buds whose scales are pointed and fringed by hairs. I treat our species in two groups—tall and low—plus Bartram's. But the hybrids cross the group lines freely.

OUR TWO CHOKEBERRIES are very similar to shads but don't have fringed scales. They vary between smooth plants with black fruits in the north and hairy ones with red fruits along the east coast. In-between plants occur in between. We illustrate the two extremes here, and leave you to decide what, if anything, you want to do with the others.

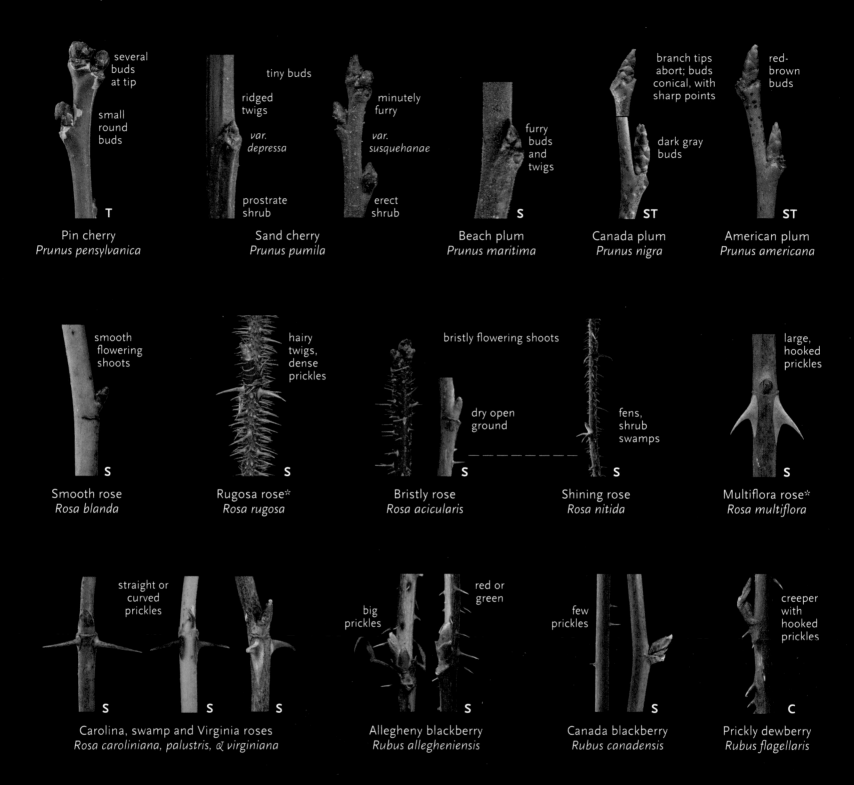

several buds at tip

small round buds

T

Pin cherry
Prunus pensylvanica

tiny buds

ridged twigs

var. depressa

prostrate shrub

Sand cherry
Prunus pumila

minutely furry

var. susquehanae

erect shrub

furry buds and twigs

S

Beach plum
Prunus maritima

branch tips abort; buds conical, with sharp points

dark gray buds

ST

Canada plum
Prunus nigra

red-brown buds

ST

American plum
Prunus americana

smooth flowering shoots

S

Smooth rose
Rosa blanda

hairy twigs, dense prickles

S

Rugosa rose*
Rosa rugosa

bristly flowering shoots

dry open ground

S

Bristly rose
Rosa acicularis

fens, shrub swamps

S

Shining rose
Rosa nitida

large, hooked prickles

S

Multiflora rose*
Rosa multiflora

straight or curved prickles

S **S** **S**

Carolina, swamp and Virginia roses
Rosa caroliniana, palustris, & virginiana

big prickles

red or green

S

Allegheny blackberry
Rubus allegheniensis

few prickles

S

Canada blackberry
Rubus canadensis

creeper with hooked prickles

C

Prickly dewberry
Rubus flagellaris

THE HAWTHORNS, p. 31, also hybridize. Few people can identify them in the summer, and probably no one in the winter. Just call them hawthorns and let it go at that.

THE CHERRIES AND PLUMS, *Prunus*, are usually all identifiable in the winter, though the differences are small. All have a cherry pit (cyanic) smell, which is stronger in some species than others. The plums have aborted twig tips along side the terminal buds; the sand cherry has the ridged twigs and often creeps, the beach plum has hairy buds, and the pin cherry clustered buds. Anything without any of these features is a black cherry if it has warm red-brown buds) or a choke cherry if it has drab gray buds with light edges to the scales.

ROSACEAE, ROSE FAMILY

——— arching stems ———

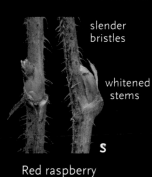

leaves often
evergreen

creeping
stems,
delicate
bristles

C

Bristly dewberry
Rubus hispidus

partly
erect,
stiffer
bristles

S

Setose blackberry
Rubus setosus

slender
bristles

whitened
stems

S

Red raspberry
Rubus idaeus

hooked
prickles

S

Black raspberry
Rubus occidentalis

shreddy
bark

no
prickles

S

Flowering raspberry
Rubus odoratus

RUTACEAE, RUE FAMILY

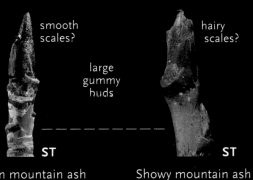

smooth
scales?

large
gummy
buds

ST

American mountain ash
Sorbus americana

hairy
scales?

ST

Showy mountain ash
Sorbus decora

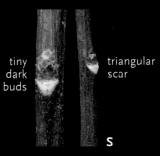

tiny
dark
buds

triangular
scar

S

Meadowsweet spiraea
Spiraea alba

small
round
buds

matted
woolly
hairs

S

Steeplebush
Spiraea tomentosa

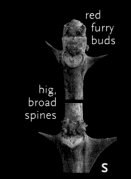

red
furry
buds

big,
broad
spines

S

Prickly ash
Zanthoxylum americanum

SALICACEAE, WILLOW FAMILY: *POPULUS,* POPLARS, ASPENS (Three bundle scars, lowest bud scale directly above the leaf scar) ———

dark,
pointed,
shiny
buds

T

Quaking aspen
Populus tremuloides

frosted
buds

T

Big-toothed aspen
Populus grandidentata

long, sharp, resinous buds

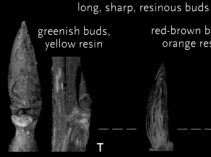

greenish buds,
yellow resin

red-brown buds,
orange resin

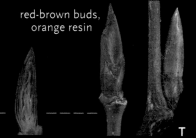

T

Cottonwood
Populus deltoides

T

Balsam poplar
Populus balsamifera

THE ROSES, *Rosa,* and the briers, *Rubus,* are variable groups with weak species lines that are difficult to identify winter or summer. The best approach, except for distinctive species (rugosa and multiflora roses, the three raspberries) is to recognize broad, arbitrary groups defined by stem and leaf characters. We give the most convenient winter characters here but do not really trust them.

THE TWO ASPENS are easily separated by their leaves (p. 51) or the surface of the buds, shiny in quaking aspen and dull and minutely hairy in big-toothed. The cottonwood and balsam poplar are best separated by their leaves. In winter the balsam poplar's buds are often slenderer and darker, but this relative, and probably shouldn't be relied upon.

33

QUICK GUIDES TO BUDS & TWIGS

QUICK GUIDES TO LEAVES

EVERGREENS

OPPOSITE BUDS

ALTERNATE BUDS

OPPOSITE LEAVES

ALTERNATE LEAVES

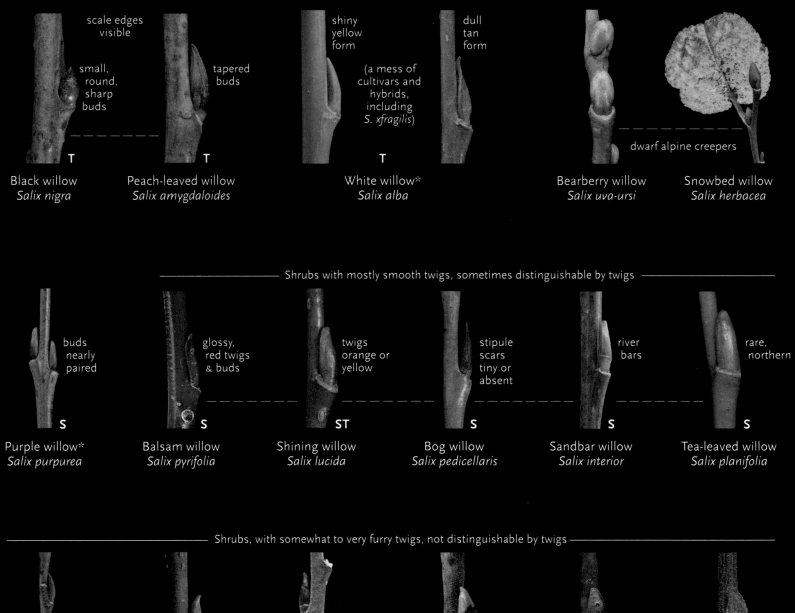

scale edges visible

small, round, sharp buds

T

tapered buds

T

shiny yellow form

(a mess of cultivars and hybrids, including *S. xfragilis*)

T

dull tan form

dwarf alpine creepers

Black willow
Salix nigra

Peach-leaved willow
Salix amygdaloides

White willow*
Salix alba

Bearberry willow
Salix uva-ursi

Snowbed willow
Salix herbacea

———————— Shrubs with mostly smooth twigs, sometimes distinguishable by twigs ————————

buds nearly paired

S

glossy, red twigs & buds

S

twigs orange or yellow

ST

stipule scars tiny or absent

S

river bars

S

rare, northern

S

Purple willow*
Salix purpurea

Balsam willow
Salix pyrifolia

Shining willow
Salix lucida

Bog willow
Salix pedicellaris

Sandbar willow
Salix interior

Tea-leaved willow
Salix planifolia

———————— Shrubs, with somewhat to very furry twigs, not distinguishable by twigs ————————

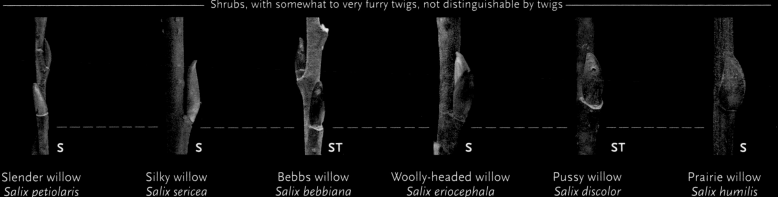

S

S

ST

S

ST

S

Slender willow
Salix petiolaris

Silky willow
Salix sericea

Bebbs willow
Salix bebbiana

Woolly-headed willow
Salix eriocephala

Pussy willow
Salix discolor

Prairie willow
Salix humilis

THE WILLOWS are very identifiable if you have leaves or fruit, but not if you only have twigs. The buds vary a lot in character and shape but not, to my eyes, in a useful or diagnostic way. Your best option is to root around and find leaves or leave them to spring. Failing that, use size and ecology. Among the trees, *nigra* has short round buds, while *alba* has long flat ones. In bogs and fens, *candida* has short buds with matted white hairs, while *pedicellaris* has smooth twigs with longer

one. On river bars, *lucida* and *exigua* have smooth twigs and *sericea* and *eriocephala* downy ones. In the boreal, *pyrifolia* has some of the reddest and glossiest twigs of any of our willows. And in everyday willow thickets, the plants with dull furry twigs are likely one of four or five common species (*discolor, sericea, bebbiana, eriocephala,* and *petiolaris*) but which one, so far as I know, can't be determined without leaves or flowers.

QUICK GUIDES TO BUDS & TWIGS

QUICK GUIDES TO LEAVES

EVERGREENS

OPPOSITE BUDS

ALTERNATE BUDS

OPPOSITE LEAVES

ALTERNATE LEAVES

SALICACEAE, WILLOW FAMILY   **SOLANACEAE,** POTATO FAMILY   **THYMELIACEAE,** MEZEREUM FAMILY

white matted woolly hairs

S

Hoary willow
Salix candida

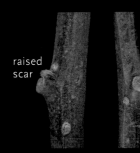

raised scar

green, ridged twigs

V

Bittersweet nightshade*
Solanum dulcamara

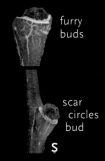

furry buds

scar circles bud

S

Leatherwood
Dirca palustris

ULMACEAE, ELM FAMILY: *ULMUS,* ELMS (Scales in 2 rows, leaf scars half-round, with 3 clear bundle scars) ───

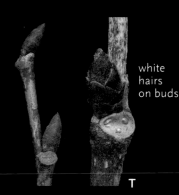

white hairs on buds

T

American elm
Ulmus americana

black buds, copper hairs mixed with white

T

Slippery elm
Ulmus rubra

corky ridges on branches

T

Cork elm
Ulmus thomasii

VITACEAE, GRAPE FAMILY (Vines climbing with branching tendrils) ───

tendrils thick, coiling; buds hairy at tips, leaf scars small   tendrils thin, buds smooth, leaf scars large and round

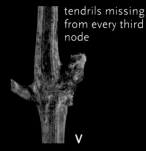

tendrils missing from every third node

V

Summer and riverbank grapes
Vitis aestivalis & riparia

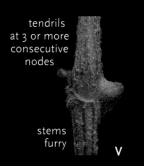

tendrils at 3 or more consecutive nodes

stems furry

V

Fox grape
Vitis labrusca

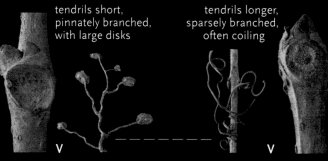

tendrils short, pinnately branched, with large disks

tendrils longer, sparsely branched, often coiling

V   V

Virginia creeper
Parthenocissus quinquefolia   Virginia creeper
Parthenocissus inserta

THE ELMS all have dark, flattened buds with the scales in two rows. Slippery elm has shiny, dark red hairs on the bud scales. Cork elm has corky ridges on the branches; common elm has neither.

THE GRAPES have thick, coiling tendrils and oval buds with a tuft of hairs at their tips. Fox grape, with tendrils opposite most of the leaf scars, is separable in the winter. Summer and riverbank grapes, with every third leaf scar missing a tendril, aren't.

THE VIRGINIA CREEPERS, *Parthenocissus,* are separated from the grapes by their thinner tendrils, conspicuous round leaf scars, and low pyramidal buds. They are told from each other by how much the tendrils branch and whether they have disks at the end. All these characters vary, and it is possible that *quinquefolia* has one sort of tendril when scrambling and another sort when climbing. This is deep water, and you may be forgiven for not wanting to venture too far into it.

Maple-leaved viburnum
Viburnum acerifolium
soft fur

Highbush cranberry
Viburnum trilobum
broad flaring lobes
glands on leafstalk

Squashberry
Viburnum edule
small lobes
glands on lower teeth

Red elderberry
Sambucus racemosa
brown pith
compound leaves, sharp fine teeth, fat twigs

Common elderberry
Sambucus canadensis
white pith

Hobblebush
Viburnum lantanoides
large, rounded, fine teeth

Arrowwood
Viburnum dentatum
few hairs
twigs ridged
wetlands

Rafinesque's viburnum
Viburnum rafinesquianum
soft fur
dry hills

Wild raisin
Viburnum cassinoides
blunt teeth

Nannyberry
Viburnum lentago
sharper teeth
winged leafstalk

VIBURNUM LEAVES are quite variable in shape, especially on young or browsed plants, and it is wise to use other characters to back up identifications. Thus the dense multi-branched hairs on the lower side are a confirmatory character for *V. acerifolium*; the petiolar glands will separate *trilobum* from *edule*; the marginal hairs and unridged twigs will separate *rafinesquianum* from *dentatum*; and the sharp teeth and winged petiole will separate *lentago* from *cassinoides*. Like leaf shape,

and in fact, all leaf characters, non of these distinctions is certain. If the leaves say one thing and the twigs and ecology something else, go with the twigs and ecology.

RED ELDERBERRY is easily separated from common elderberry is brown pith and smellier twigs. The leaves of both are variable, and probably not diagnostic.

CAPRIFOLIACEAE, HONEYSUCKLE FAMILY (Opposite leaves, mostly without teeth) ————————————

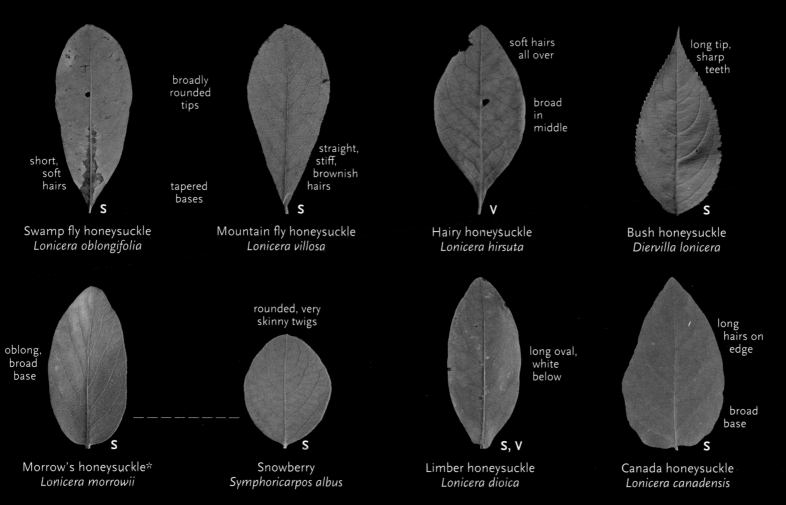

broadly
rounded
tips

short,
soft
hairs

straight,
stiff,
brownish
hairs

tapered
bases

S

Swamp fly honeysuckle
Lonicera oblongifolia

S

Mountain fly honeysuckle
Lonicera villosa

soft hairs
all over

broad
in
middle

V

Hairy honeysuckle
Lonicera hirsuta

long tip,
sharp
teeth

S

Bush honeysuckle
Diervilla lonicera

oblong,
broad
base

rounded, very
skinny twigs

S

Morrow's honeysuckle*
Lonicera morrowii

S

Snowberry
Symphoricarpos albus

long oval,
white
below

S, V

Limber honeysuckle
Lonicera dioica

long
hairs on
edge

broad
base

S

Canada honeysuckle
Lonicera canadensis

CORNACEAE, DOGWOOD FAMILY: *CORNUS*, DOGWOODS (Opposite leaves, no teeth, veins curve way forward, duck-bill buds)

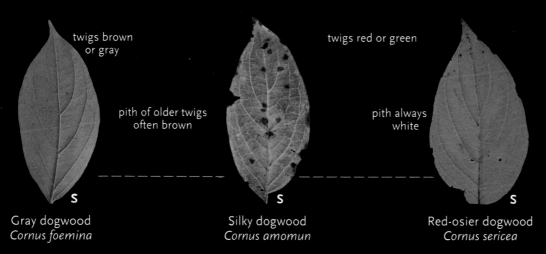

twigs brown
or gray

pith of older twigs
often brown

S

Gray dogwood
Cornus foemina

twigs red or green

pith always
white

S

Silky dogwood
Cornus amomun

S

Red-osier dogwood
Cornus sericea

TRUE HONEYSUCKLES, *Lonicera*, have blunt leaves and no teeth at all. Anything with teeth or a sharp point is something else. The native snowberry, which can resemble small plants of the alien Morrow's honeysuckle, is a low plant of dry rocky or sandy ground with very slender, dark brown twigs and thin, rounded, leaves. Morrow's honeysuckle is a large, aggressive, ecological generalist that typically has oblong leaves and thicker, lighter-colored leaves.

DOGWOODS have toothless oval leaves with acuminate tip and veins that curve forwards. The three narrow-leaved dogwoods— *foemina, amomum,* and *sericea*—are quite similar, though *sericea* tends to have more veins. Use the twig and pith color, remembering that the red-twig dogwoods turn green in summer, and the red fruit stalks of *foemina* and *sericea* for confirmation.

usually rough above

twigs with purple streaks

broad base

S

Round-leaved dogwood
Cornus rugosa

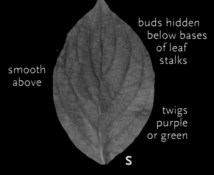

buds hidden below bases of leaf stalks

smooth above

twigs purple or green

S

Flowering dogwood
Cornus florida

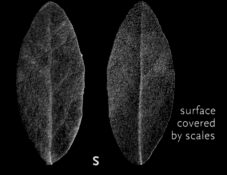

surface covered by scales

S

Buffaloberry
Shepherdia canadensis

OLEACEAE, OLIVE FAMILY: FRAXINUS, ASHES (Opposite leaves with teeth, sometimes compound or lobed) ————————

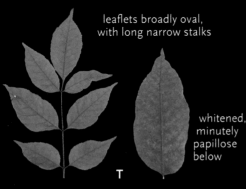

leaflets broadly oval, with long narrow stalks

whitened, minutely papillose below

T

White ash
Fraxinus americana

leaflets narrower, with short, winged stalks

stalks often furry

T

Red ash
Fraxinus pennsylvanica

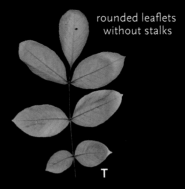

rounded leaflets without stalks

T

Black ash
Fraxinus nigra

RANUNCULACEAE, BUTTERCUP FAMILY: *CLEMATIS* (Vines with opposite compound leaves)   **RUBIACEAE,** MADDER FAMILY

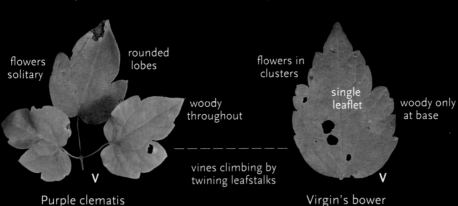

flowers solitary

rounded lobes

woody throughout

V

Purple clematis
Clematis occidentalis

—————— vines climbing by twining leafstalks

flowers in clusters

single leaflet

woody only at base

V

Virgin's bower
Clematis virginiana

large glossy, oval leaves without teeth

S

Buttonbush
Cephalanthus occidentalis

ASH LEAVES are similar; familiarity helps. The long stalks of the leaflets are a traditional character for white ash but not a particularly good one. The whitening, caused by minute papillae visible at 30×, is a better one. The winged leaflet stalks are a good red ash character; furry leafstalks are a nice confirmation but not always there. Black ash leaflets have almost no stalks at all, and are often broadly oval; the crumbly bark, without braiding ridges, is a good confirmation.

CLEMATISES are climbing vines with compound leaves, often dying back in winter. The best distinctions are ecology and flowers. Purple clematis grows in fertile rocky woods and has a few large, long-stalked flowers. Virgin's bower grows in wet thickets, and has many small, short-stalked flowers in clusters.

SAPINDACEAE, SOAPBERRY FAMILY: *ACER,* MAPLES (Trees with opposite, palmately lobed or compound leaves)

QUICK GUIDES TO BUDS & TWIGS

QUICK GUIDES TO LEAVES

EVERGREENS

OPPOSITE BUDS

ALTERNATE BUDS

OPPOSITE LEAVES

ALTERNATE LEAVES

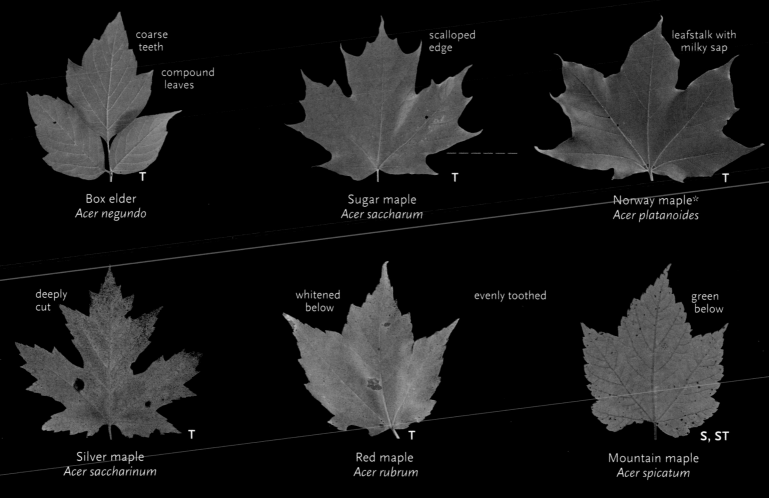

coarse teeth

compound leaves

T

Box elder
Acer negundo

scalloped edge

T

Sugar maple
Acer saccharum

leafstalk with milky sap

T

Norway maple*
Acer platanoides

deeply cut

T

Silver maple
Acer saccharinum

whitened below

evenly toothed

T

Red maple
Acer rubrum

green below

S, ST

Mountain maple
Acer spicatum

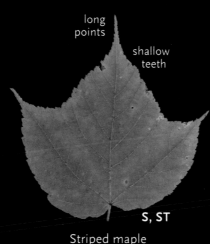

long points

shallow teeth

S, ST

Striped maple
Acer pensylvanicum

STAPHYLEACEAE, BLADDERNUT FAMILY

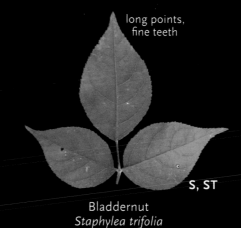

long points, fine teeth

S, ST

Bladdernut
Staphylea trifolia

MAPLE LEAVES are easy. Note the coarse teeth on the box elder, the deep divisions of silver maple, the whitened undersides of red maple, and the rounded outline and long lobe tips of striped maple. Sugar maple is similar to the introduced Norway maple, and in disturbed woods where both may occur, the deeply grooved bark and squat buds, p. 25, of Norway maple will help. Silver maples and red maples interbreed, and occasionally you will encounter trees whose leaves are intermediate between the two.

BLADDERNUT is a tall shrub, found in colonies on river banks and near ledges in rocky fertile woods. It is our only woody plant with finely toothed, opposite, trifoliate leaves. Box elder is similar but has coarse teeth. Fragrant sumac and poison ivy, p. 40, are alternate and have coarse teeth.

39

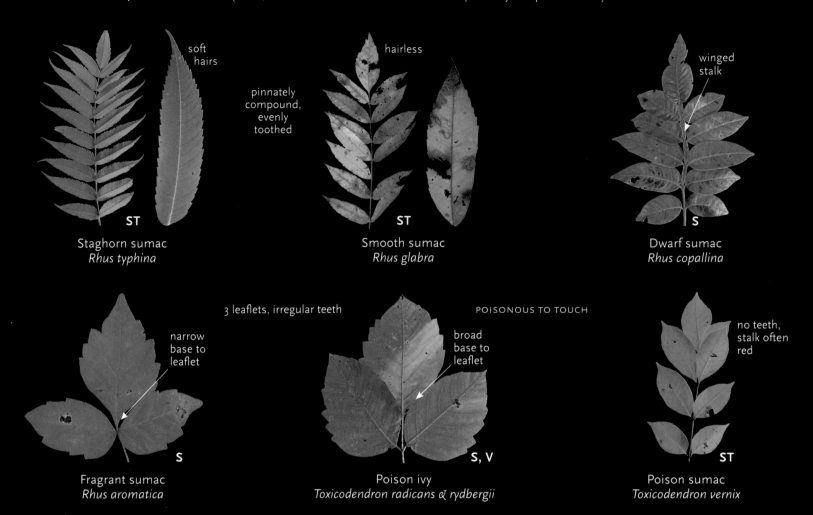

Staghorn sumac
Rhus typhina

soft hairs

pinnately compound, evenly toothed

Smooth sumac
Rhus glabra

hairless

Dwarf sumac
Rhus copallina

winged stalk

Fragrant sumac
Rhus aromatica

3 leaflets, irregular teeth

narrow base to leaflet

Poison ivy
Toxicodendron radicans & rydbergii

POISONOUS TO TOUCH

broad base to leaflet

Poison sumac
Toxicodendron vernix

no teeth, stalk often red

AQUIFOLIACEAE, HOLLY FAMILY

BERBERIDACEAE, BARBERRY FAMILY

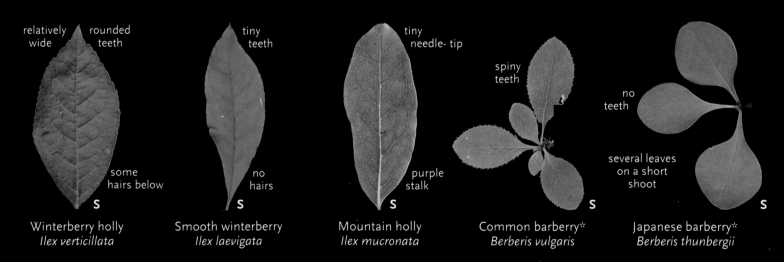

Winterberry holly
Ilex verticillata

relatively wide | rounded teeth

some hairs below

Smooth winterberry
Ilex laevigata

tiny teeth

no hairs

Mountain holly
Ilex mucronata

tiny needle- tip

purple stalk

Common barberry*
Berberis vulgaris

spiny teeth

Japanese barberry*
Berberis thunbergii

no teeth

several leaves on a short shoot

FRAGRANT SUMAC, an uncommon species of dry, fertile, rocky hills, much resembles poison ivy but is an erect, branchy shrub and has a tapering base to the terminal leaflet; our poison ivies mostly creep or climb and have rounded bases to the terminal leaflets. The separation of the two poison ivies is mostly based on growth form: *radicans* climbs trees by rootlets or (mostly south of us) can grow as a shrub; *rydbergii* just creeps. Also note that smooth sumac is basically a hairless version of staghorn, and intermediates are common.

THE TWO WINTERBERRIES are close: *verticillata* is much commoner and has broader, toothier leaves that usually have some hairs on the lower sides. *Laevigata* is southern and has narrower, smoother leaves with only tiny teeth.

BETULACEAE, BIRCH FAMILY (Trees and shrubs with oval, toothed leaves; often two sizes of teeth) ───────────

QUICK GUIDES TO
BUDS & TWIGS

QUICK GUIDES
TO LEAVES

EVERGREENS

OPPOSITE BUDS

ALTERNATE BUDS

OPPOSITE LEAVES

ALTERNATE LEAVES

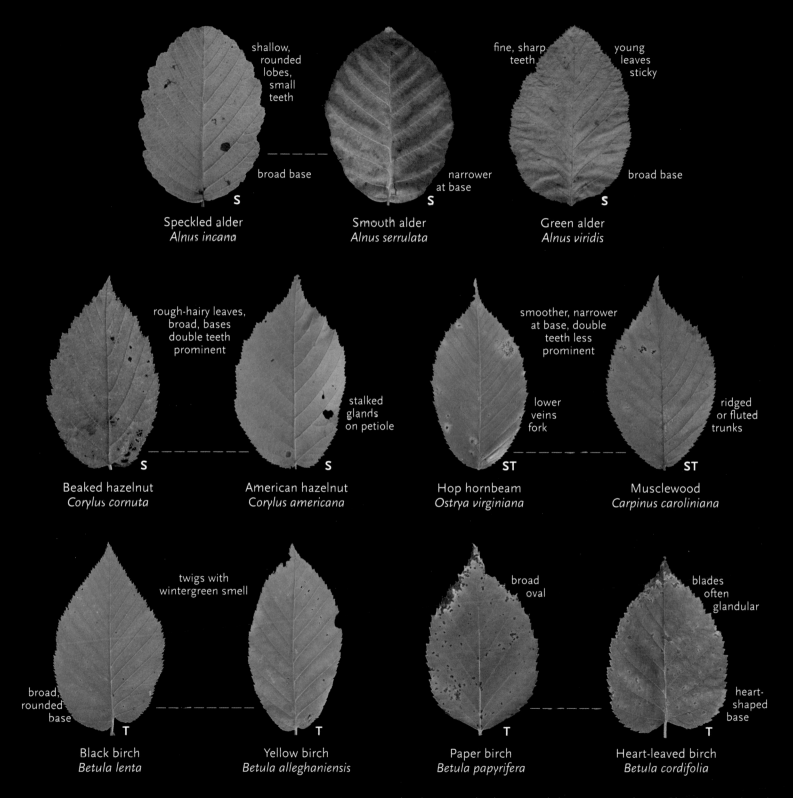

shallow, rounded lobes, small teeth

broad base

fine, sharp teeth

young leaves sticky

narrower at base

broad base

S
Speckled alder
Alnus incana

S
Smooth alder
Alnus serrulata

S
Green alder
Alnus viridis

rough-hairy leaves, broad, bases double teeth prominent

stalked glands on petiole

smoother, narrower at base, double teeth less prominent

lower veins fork

ridged or fluted trunks

S
Beaked hazelnut
Corylus cornuta

S
American hazelnut
Corylus americana

ST
Hop hornbeam
Ostrya virginiana

ST
Musclewood
Carpinus caroliniana

twigs with wintergreen smell

broad oval

blades often glandular

broad, rounded base

heart-shaped base

T
Black birch
Betula lenta

T
Yellow birch
Betula alleghaniensis

T
Paper birch
Betula papyrifera

T
Heart-leaved birch
Betula cordifolia

BIRCH FAMILY LEAVES are commonly oval or rounded triangular and weakly or strongly double toothed, meaning that small teeth alternate with or ride on the backs of larger ones. Recognizing the species from the leaves alone can be hard. The alders (large oval leaves, acute tips) and the hazelnuts (hairy, broad-based and broad-shouldered leaves with coarse double teeth and acuminate tips) are easy. Gray birch (triangular leaves with long tips) and paper birch (broad oval leaves with relatively coarse teeth) are usually identifiable, though the separation is (much) blurred northwards where the heart-leaved birch and its hybrids occur. Musclewood, hop hornbeam, black birch, and yellow birch are similar and variable. Use the leaves when you can, but be prepared to rely on bark, buds, and fruit when you can't.

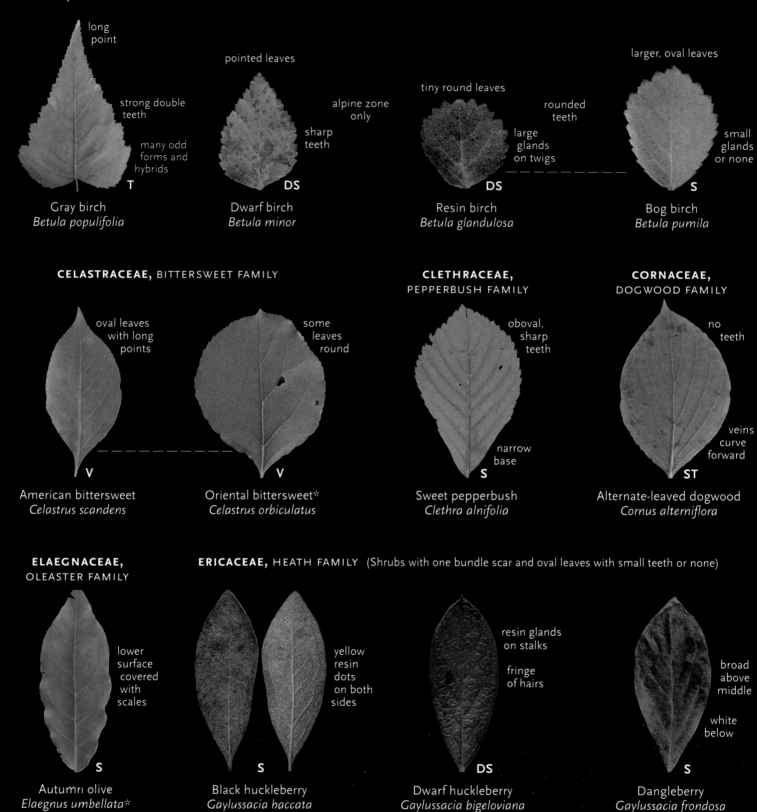

long
point

strong double
teeth

many odd
forms and
hybrids

T

Gray birch
Betula populifolia

pointed leaves

alpine zone
only

sharp
teeth

DS

Dwarf birch
Betula minor

tiny round leaves

rounded
teeth

large
glands
on twigs

DS

Resin birch
Betula glandulosa

larger, oval leaves

small
glands
or none

S

Bog birch
Betula pumila

CELASTRACEAE, BITTERSWEET FAMILY

oval leaves
with long
points

V

American bittersweet
Celastrus scandens

some
leaves
round

V

Oriental bittersweet⁎
Celastrus orbiculatus

CLETHRACEAE, PEPPERBUSH FAMILY

oboval,
sharp
teeth

narrow
base

S

Sweet pepperbush
Clethra alnifolia

CORNACEAE, DOGWOOD FAMILY

no
teeth

veins
curve
forward

ST

Alternate-leaved dogwood
Cornus alterniflora

ELAEGNACEAE, OLEASTER FAMILY

lower
surface
covered
with
scales

S

Autumn olive
Elaegnus umbellata⁎

ERICACEAE, HEATH FAMILY (Shrubs with one bundle scar and oval leaves with small teeth or none)

yellow
resin
dots
on both
sides

S

Black huckleberry
Gaylussacia baccata

resin glands
on stalks

fringe
of hairs

DS

Dwarf huckleberry
Gaylussacia bigeloviana

broad
above
middle

white
below

S

Dangleberry
Gaylussacia frondosa

THE SHRUBBY BIRCHES—dwarf birch, resin birch, and bog birch—are interrelated and difficult. The best characters are the pointed tips and weak double-toothing of the dwarf birch, the small size and alpine habitat of the resin birch, and the larger size and fen habitat of the bog birch. The resin birch has very glandular twigs; the other two can be glandular or not as they choose.

BLACK HUCKLEBERRY AND DWARF HUCKLEBERRY grow side by side in the bogs of the north Atlantic coast. Small plants of black huckleberry are very similar to the dwarf huckleberry. I look for the stalked yellow glands (which look like minute lines rather than dots) on the upper surface of dwarf huckleberry, plus a fringe of curved white hairs on the leaf edge, and, in fall, a slightly brighter or more orange red color.

ERICACEAE, HEATH FAMILY

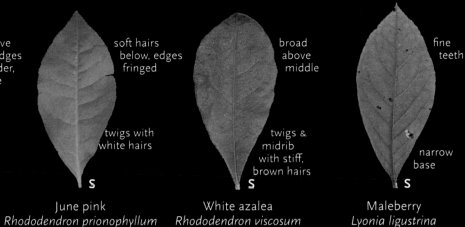

rounded teeth, showy veins

alpine creeper

Alpine bearberry
Arctous alpina

broad above middle, edges rolled under, often pale below

sparse, stiff hairs

S

Rhodora
Rhododendron canadense

soft hairs below, edges fringed

twigs with white hairs

S

June pink
Rhododendron prionophyllum

broad above middle

twigs & midrib with stiff, brown hairs

S

White azalea
Rhododendron viscosum

fine teeth

narrow base

S

Maleberry
Lyonia ligustrina

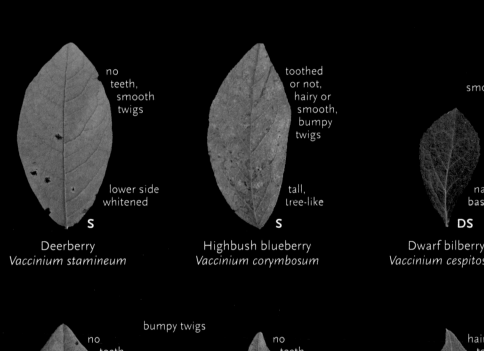

no teeth, smooth twigs

lower side whitened

S

Deerberry
Vaccinium stamineum

toothed or not, hairy or smooth, bumpy twigs

tall, tree-like

S

Highbush blueberry
Vaccinium corymbosum

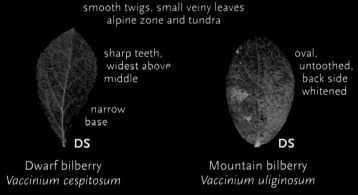

smooth twigs, small veiny leaves
alpine zone and tundra

sharp teeth, widest above middle

narrow base

DS

Dwarf bilberry
Vaccinium cespitosum

oval, untoothed, back side whitened

DS

Mountain bilberry
Vaccinium uliginosum

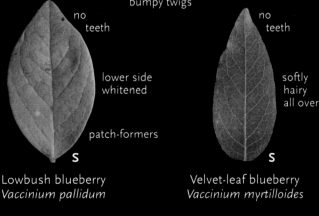

bumpy twigs

no teeth

lower side whitened

patch-formers

S

Lowbush blueberry
Vaccinium pallidum

no teeth

softly hairy all over

S

Velvet-leaf blueberry
Vaccinium myrtilloides

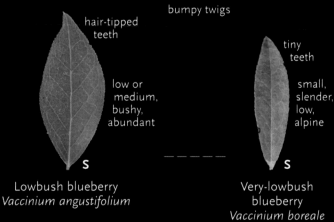

hair-tipped teeth

low or medium, bushy, abundant

S

Lowbush blueberry
Vaccinium angustifolium

bumpy twigs

tiny teeth

small, slender, low, alpine

S

Very-lowbush blueberry
Vaccinium boreale

THE DECIDUOUS RHODODENDRONS have leaves with conspicuous hairs along the edges that tend to cluster at the twig tips and appear whorled. Rhodora is a northern mountain and bog plant with whitened twigs and distinctive, light blue-green leaves. June pink and white azalea are more southern, with broader, darker green leaves. They differ, seemingly reliably, in leaf shape and in the types of hairs.

THE BLUEBERRIES and their close relatives are variable and difficult, always hard to separate by their leaves and sometimes hard to separate at all. The small bumps on the twigs and sharp-pointed bud scales will separate the blueberries as a group. Within the group, the size of the plants, size and shape of the leaves, presence of teeth, and whitening of the leaf underside are the most useful characters.

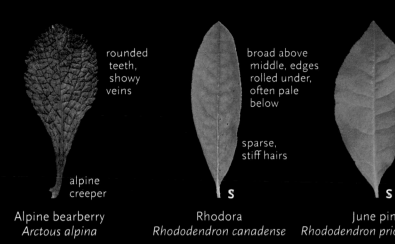

QUICK GUIDES TO BUDS & TWIGS

QUICK GUIDES TO LEAVES

EVERGREENS

OPPOSITE BUDS

ALTERNATE BUDS

OPPOSITE LEAVES

ALTERNATE LEAVES

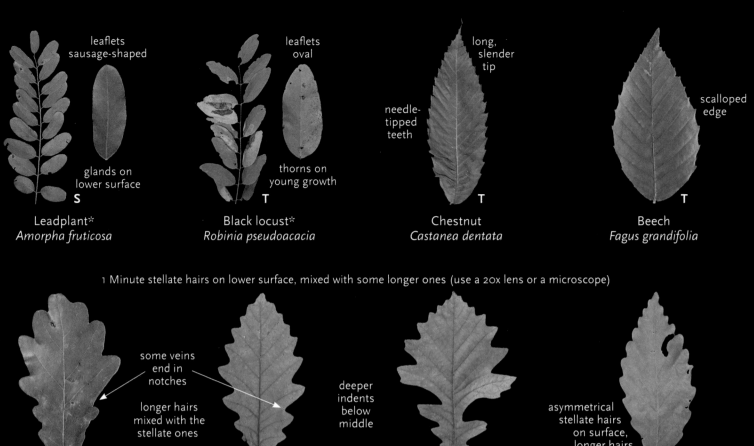

leaflets
sausage-shaped

glands on
lower surface

S

Leadplant*
Amorpha fruticosa

leaflets
oval

thorns on
young growth

T

Black locust*
Robinia pseudoacacia

long,
slender
tip

needle-
tipped
teeth

T

Chestnut
Castanea dentata

scalloped
edge

T

Beech
Fagus grandifolia

1 Minute stellate hairs on lower surface, mixed with some longer ones (use a 20x lens or a microscope)

some veins
end in
notches

longer hairs
mixed with the
stellate ones

T

Swamp white oak
Quercus bicolor

deeper
indents
below
middle

T

Bur oak
Quercus macrocarpa

asymmetrical
stellate hairs
on surface,
longer hairs
in vein axils

T

Chestnut oak
Quercus montana

2 Minute stellate hairs on lower surface

lobes sharp
or rounded,
no long hairs

T

Yellow oak
Quercus muehlenbergia

colonial
shrub

Dwarf chestnut oak
Quercus prinoides

3 Two distinctive species

rounded
lobes

smooth &
whitened
below

T

White oak
Quercus alba

blunt lobes with
needle tips

densely furry
below

ST

Scrub oak
Quercus ilicifolia

OAK LEAVES vary a lot in shape, even within the same species. For identification, the hairs on the lower side, observed with a 20x lens, are very useful. The first five species shown here can resemble each other in leave shape, and all have the lower leaf surfaces covered by minute stellate (multibranched) hairs. The swamp white oak and chestnut oak have longer hairs as well. The others do not.

White oak and scrub oak are distinctive species. The white oak, a large tree, has long rounded lobes and a whitened lower surface. The scrub oak, a multi-stemmed colonial shrub or small tree, has needle-tipped, bluntly triangular lobes, and a dense, felty covering of hairs below. Yellow oak and dwarf chestnut oak, are closely related, and differ mostly in growth habit. Yellow oak is a tree, dwarf chestnut a low, multi-stemmed shrub that may flower when only a few feet high.

FAGACEAE, BEECH FAMILY

3 Bristle-tipped lobes; leaf shape extremely variable, not reliable for identification

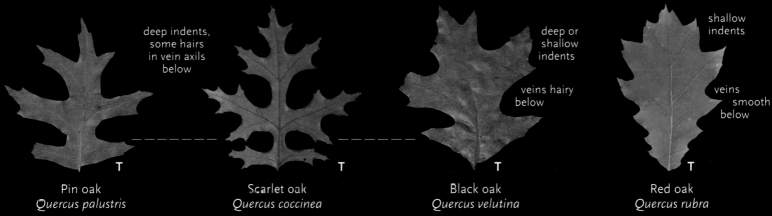

deep indents, some hairs in vein axils below

deep or shallow indents

veins hairy below

shallow indents

veins smooth below

| Pin oak | Scarlet oak | Black oak | Red oak |
| *Quercus palustris* | *Quercus coccinea* | *Quercus velutina* | *Quercus rubra* |

GROSSULARIACEAE, GOOSEBERRY FAMILY: *RIBES*, CURRANTS & GOOSEBERRIES (Shrubs, often spiny, with palmately lobed leaves)

softly hairy, fruits with spines

sparsely hairy, fruits smooth, lobes more pointed, bases rounded or v-shaped

heart-shaped base

Prickly gooseberry
Ribes cynosbati

Round-leaved gooseberry
Ribes rotundifolium

Bristly gooseberry
Ribes hirtellum

small, dark buds

low shrubs, often partly creeping

large red or green buds, musty smell

deep indents, bristly yellow twigs

yellow resin glands

Swamp red currant
Ribes triste

Skunk currant
Ribes glandulosa

Swamp black currant
Ribes lacustre

American currant
Ribes americana

THE RED OAK AND ITS RELATIVES all have leaves with bristle tips and sharp lobes and small amounts of fur in the vein axils below. The red oak is usually more shallowly lobed than the other three and the black oak more hairy along the veins, but these characters all vary. The best characters, and sometimes the only good characters, are in the buds and the acorns.

OUR GOOSEBERRIES are close and best distinguished by their flowers. The leaves differ, with *cynosbati* having the roundest base and *routundifolium* the narrowest, but they all overlap. The currants are easier: American has glandular leaves, swamp black has bristly yellow twigs, and skunk current a pungent musty smell. All of them vary, and it is wise to check the buds to confirm the ID.

QUICK GUIDES TO BUDS & TWIGS

QUICK GUIDES TO LEAVES

EVERGREENS

OPPOSITE BUDS

ALTERNATE BUDS

OPPOSITE LEAVES

ALTERNATE

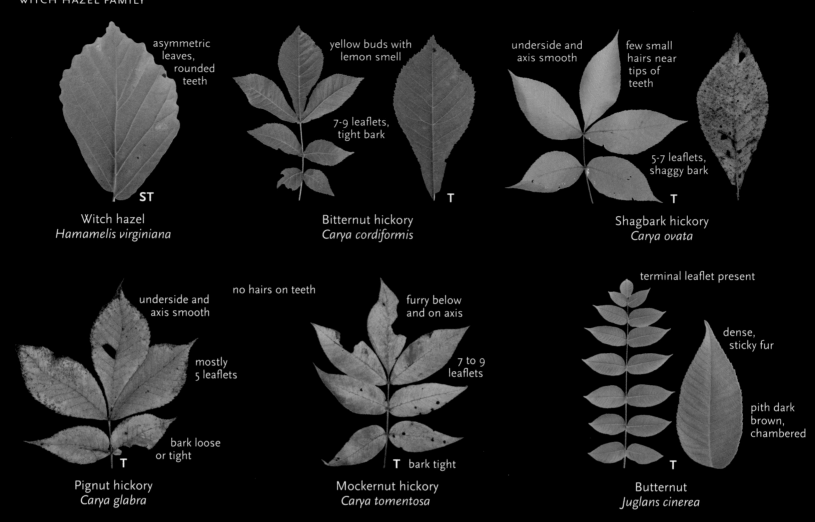

asymmetric leaves, rounded teeth

ST

Witch hazel
Hamamelis virginiana

yellow buds with lemon smell

7-9 leaflets, tight bark

T

Bitternut hickory
Carya cordiformis

underside and axis smooth

few small hairs near tips of teeth

5-7 leaflets, shaggy bark

T

Shagbark hickory
Carya ovata

underside and axis smooth

no hairs on teeth

mostly 5 leaflets

bark loose or tight

T

Pignut hickory
Carya glabra

furry below and on axis

7 to 9 leaflets

T bark tight

Mockernut hickory
Carya tomentosa

terminal leaflet present

dense, sticky fur

pith dark brown, chambered

T

Butternut
Juglans cinerea

LAURACEAE, LAUREL FAMILY (Trees and shrubs with green, spicy-scented twigs and untoothed leaves.)

MALVACEAE, MALLOW FAMILY

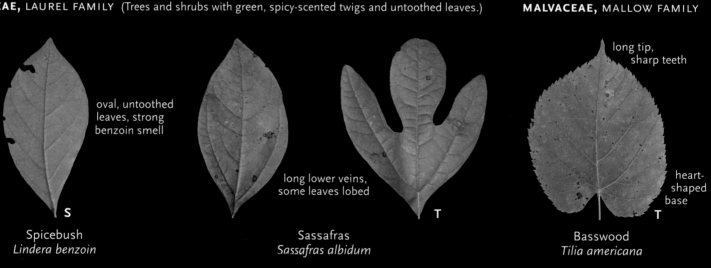

oval, untoothed leaves, strong benzoin smell

S

Spicebush
Lindera benzoin

long lower veins, some leaves lobed

T

Sassafras
Sassafras albidum

long tip, sharp teeth

heart-shaped base

T

Basswood
Tilia americana

HICKORY LEAVES resemble ashes (p. 38) but are alternately arranged and have sharper and more even teeth. The species are fairly easy to distinguish. Bitternut, common in fertile woods has yellow scales on the buds (p. 30), young twigs, and leafstalks. Its buds smell like lemon oil when rubbed. Mockernut, a southern species that reaches the southern edge of the Northern Forest region, has rounded buds and dense fur on the underside of the leaves. Pignut and shagbark, the commonest hickories on dry rocky hills, often occur together. They have leaves that are mostly hairless below, and oval buds. Shagbark, which extends farther north, has loose bark, larger fruits with thick husks, and tiny tufts of hair on the leaf margins. It is close to pignut, and intermediate trees with partly shaggy bark are common.

MENISPERMACEAE
MOONSEED FAMILY

MORACEAE, MULBERRY FAMILY

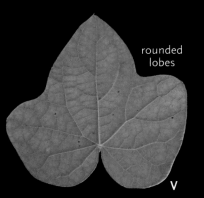

rounded lobes

V

Moonseed
Menispermum canadense

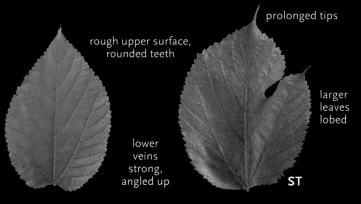

rough upper surface, rounded teeth

prolonged tips

larger leaves lobed

lower veins strong, angled up

ST

Red mulberry
Morus rubra

small, sharp teeth

lopsided base

T

Hackberry
Celtis laevigata

MYRICACEAE, BAYBERRY FAMILY (Aromatic, yellow resin dots)

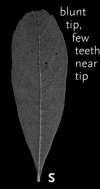

blunt tip, few teeth near tip

S

Sweet gale
Myrica gale

pointed tip

S

Bayberry
Myrica pensylvanica

pinnately lobed

S

Sweet fern
Comptonia peregrina

NYSSACEAE, TUPELO FAMILY

broad oval, no teeth

tapered base

T

Black gum
Nyssa sylvatica

PLATANACEAE, PLANE-TREE FAMILY

sharp lobes

no teeth

T

Sycamore
Platanus occidentalis

PINACEAE, PINE FAMILY

needles soft, blue-green, in tufts on older branches

T

Tamarack
Larix laricina

RHAMNACEAE, BUCKTHORN FAMILY (lowest veins bend forwards, small teeth or none)

leaves all alternate

rounded teeth

low shrub, < 1m high

Alder-leaved buckthorn
Rhamnus alnifolia

some leaves almost paired

ST

Common buckthorn*
Rhamnus cathartica

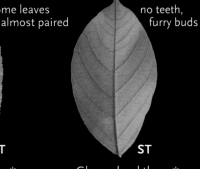

no teeth, furry buds

ST

Glossy buckthorn*
Frangula alnus

triangular leaves, strong lower veins

S

New Jersey tea
Ceanothus americanus

THE BUCKTHORNS have veins that curve forward like a dogwood, but the leaves are at least partly alternate and often have teeth. Alder-leaved buckthorn, a low native shrub of limy wetlands and ledges, has broadly oval leaves with rounded teeth. Common buckthorn, a hedge plant that has become highly invasive on fertile soils, has oblong leaves with broad bases that vary from alternate to nearly opposite, and twigs that often end in spines. Glossy buckthorn, another impressive invader, has broadly oval leaves without teeth and furry buds. New Jersey tea, a low native shrub of dry sandy soils, has triangular leaves with low teeth and a strong lower veins. Its western relative, *Ceanothus herbaceus*, found in sandy glades and prairies, has slender oval leaves with a tapering base.

QUICK GUIDES TO BUDS & TWIGS

QUICK GUIDES TO LEAVES

EVERGREENS

OPPOSITE BUDS

ALTERNATE BUDS

OPPOSITE LEAVES

ALTERNATE LEAVES

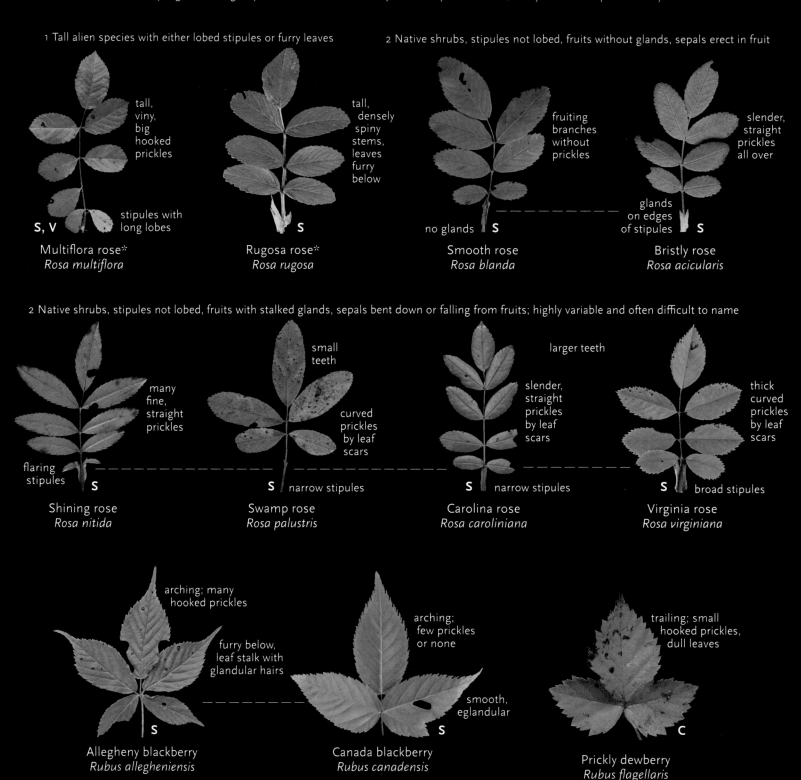

1 Tall alien species with either lobed stipules or furry leaves

tall, viny, big hooked prickles

stipules with long lobes

S, V

Multiflora rose*
Rosa multiflora

tall, densely spiny stems, leaves furry below

S

Rugosa rose*
Rosa rugosa

2 Native shrubs, stipules not lobed, fruits without glands, sepals erect in fruit

fruiting branches without prickles

no glands **S**

Smooth rose
Rosa blanda

slender, straight prickles all over

glands on edges of stipules **S**

Bristly rose
Rosa acicularis

2 Native shrubs, stipules not lobed, fruits with stalked glands, sepals bent down or falling from fruits; highly variable and often difficult to name

many fine, straight prickles

flaring stipules **S**

Shining rose
Rosa nitida

small teeth

curved prickles by leaf scars

S narrow stipules

Swamp rose
Rosa palustris

larger teeth

slender, straight prickles by leaf scars

S narrow stipules

Carolina rose
Rosa caroliniana

thick curved prickles by leaf scars

S broad stipules

Virginia rose
Rosa virginiana

arching; many hooked prickles

furry below, leaf stalk with glandular hairs

S

Allegheny blackberry
Rubus allegheniensis

arching; few prickles or none

smooth, eglandular

S

Canada blackberry
Rubus canadensis

trailing; small hooked prickles, dull leaves

C

Prickly dewberry
Rubus flagellaris

THE ROSES are our largest family of woody plants. Because of several genera (*Amelanchier, Crataegus, Rosa, Rubus*) in which hybridization and asexual reproduction are common, they are also our most confusing. They may be treated broadly, by placing the species into generalized groups, or narrowly by naming as many character combinations as you can. Neither way really works. If you use broad groups you end up putting things that don't look that much alike in the same group. If you use narrow ones, you end up with plants—often a lot of plants—that don't fit any of your names. Current practice is to treat the brambles broadly, the hawthorns narrowly, and the roses and shads in between. Authors vary, and one book's species are another one's minor variants. These are rapidly evolving groups, and all ways of dividing a rapidly evolving group into species are arbitrary.

seems creeping,
soft, slender bristles,
glossy, evergreen
leaves with
3 leaflets

C

Bristly dewberry
Rubus hispidus

stems arching,
soft, slender
bristles,
nonevergreen
leaves
with 3 or 5
leaflets

S

Setose blackberry
Rubus setosus

arching stems,
loose, papery
bark

S

Flowering raspberry
Rubus odoratus

arching stems,
3 or 5 pinnate leaflets,
whitened below

hooked
prickles

S

Black raspberry
Rubus occidentalis

soft
bristles

S

Red raspberry
Rubus idaeus

untoothed, long-hairy,
pinnately compound

S

Shrubby cinquefoil
Dasiphora fruticosa

leaflets
relatively
narrow

ST

American mountain ash
Sorbus americana

leaflets
wider

ST

Showy mountain ash
Sorbus decora

THE RASPBERRIES, SHRUBBY CINQUEFOIL, AND TWO ALIEN ROSES are distinct. The blackberries and dewberries vary a lot but can be divided into the broad groups shown here without too much trouble. The roses, on the other hand, are a lot of trouble, and some knowledge of their genetics and geography helps. Smooth rose and shining rose are northern diploids, and swamp rose a wide-ranging eastern diploid. When seen in pure form, all three are reasonably distinct.

Bristly rose is a circumboreal tetraploid or hexaploid; our plants are also reasonably distinct. Carolina rose, wide-ranging, and Virginia rose, a coastal species, are tetraploid hybrids containing different mixtures of the local diploids and, apparently, of each other. They are hard to tell from their parents and often harder to tell from each other. Both both backcross with their parents. The result is a lively and often unnameable genetic stew.

QUICK GUIDES TO
BUDS & TWIGS

QUICK GUIDES
TO LEAVES

EVERGREENS

OPPOSITE BUDS

ALTERNATE BUDS

OPPOSITE LEAVES

ALTERNATE LEAVES

low, rounded teeth

brown hairs on midrib in back

T

Black cherry
Prunus serotina

long tip, rounded teeth

widest below middle

T

Pin cherry
Prunus pensylvanica

sharp teeth

broadest above middle

ST

Choke cherry
Prunus virginiana

blunt tip

oblong leaves, low teeth, furry below

colonial shrub

Beach plum
Prunus maritima

slender leaves

narrow base

var. depressa
C

var. susquehanae
S

Sand cherry
Prunus pumila

long tips

sharp teeth

colonial

ST

American plum
Prunus americana

rounded, gland-tipped teeth

ST

Canada plum
Prunus nigra

rounded, blunt teeth, furry

ST

Apple*
Malus pumila

rounded lobes

S

Ninebark
Physocarpus opulifolius

jagged teeth, incised or lobed

ST

Hawthorns
Crataegus

woolly when young, sharp, fine teeth, rounded bases

oval or oblong

ST

Downy shadbush
Amelanchier arborea

widest above middle

ST

Canada shadbush
Amelanchier canadensis

MORE ROSE FAMILY SPECIES, and more problems. The hawthorns are a large group, treated differently by different authors. They are found mostly in pastures and second-growth, and, as pastures disappear, many of the forms have become more museum curiosities than living problems. I don't attempt to treat them here.

THE SHADS, on the other hand, are very much an up-to-date problem. The smooth and downy shads are common trees, best separated by the flowers or young foliage. Canada shad, a tree or bush, differs from downy shad in its shorter petals and leaf shape. Bartram's shad, a northern species with narrow leaves and fine sharp teeth, is distinct when pure but prone to hybridize. The low shadbushes, recognized by shrubby habit, short round leaves, and woolly ovaries, are a complex group with either weak species lines or none at all.

ROSACEAE, ROSE FAMILY

fine, sharp teeth

smooth, red when young

heart-shaped base

ST

Smooth shadbush
Amelanchier laevis

rounded leaves, coarse or fine teeth

often colonial

S

Low shadbushes
Amelanchier spicata group

fine teeth

narrow base

S

Bartram's shadbush
Amelanchier bartramiana

rounded teeth, woolly below

S

Steeplebush
Spiraea tomentosa

RUTACEAE, RUE FAMILY

sharp teeth

smooth

S

Meadowsweet spiraea
Spiraea alba

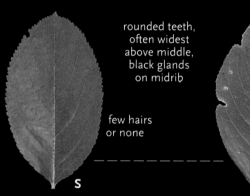

rounded teeth, often widest above middle, black glands on midrib

few hairs or none

S

Black chokeberry
Aronia melanocarpa

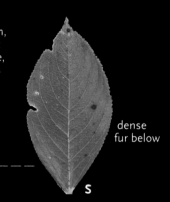

dense fur below

S

Red chokeberry
Aronia arbutifolia

pointed leaflets

S

Prickly ash
Zanthoxylum americanum

SALICACEAE, WILLOW FAMILY: *POPULUS*, POPLARS, ASPENS, COTTONWOODS (Broadly oval to triangular leaves with rounded teeth)

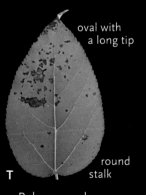

oval with a long tip

round stalk

T

Balsam poplar
Populus balsamifera

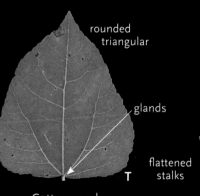

rounded triangular

glands

flattened stalks

T

Cottonwood
Populus deltoides

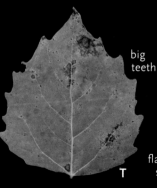

big teeth

flattened stalks

T

Big-toothed aspen
Populus grandidentata

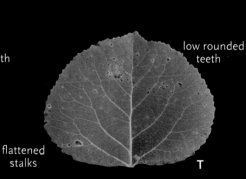

low rounded teeth

T

Quaking aspen
Populus tremuloides

THE BLACK CHOKEBERRY is small, dark-fruited, and sparsely hairy. It is found in widely in the north and in the mountains southward. The red chokeberry is taller, velvety, red-fruited, and mostly coastal. Thy hybridize, producing "purple" chokeberries, in an extensive zone where the two overlap.

THE ASPENS AND POPLARS, though hybridizable in cultivation, are distinct in the wild. The balsam poplar, a northern riverbank species, has oval leaves, a round leafstalk, and resin stains on the leaves. The big-toothed aspen, a species of dry, moderately fertile woods, has big teeth. The quaking aspen, a northern disturbance species, has rounded-triangular leaves and low teeth; the cottonwood, a wide-ranging riparian species, has a rounded triagular leaf and large glands where the stalk meets the blade.

QUICK GUIDES TO BUDS & TWIGS

QUICK GUIDES TO LEAVES

EVERGREENS

OPPOSITE BUDS

ALTERNATE BUDS

OPPOSITE LEAVES

ALTERNATE LEAVES

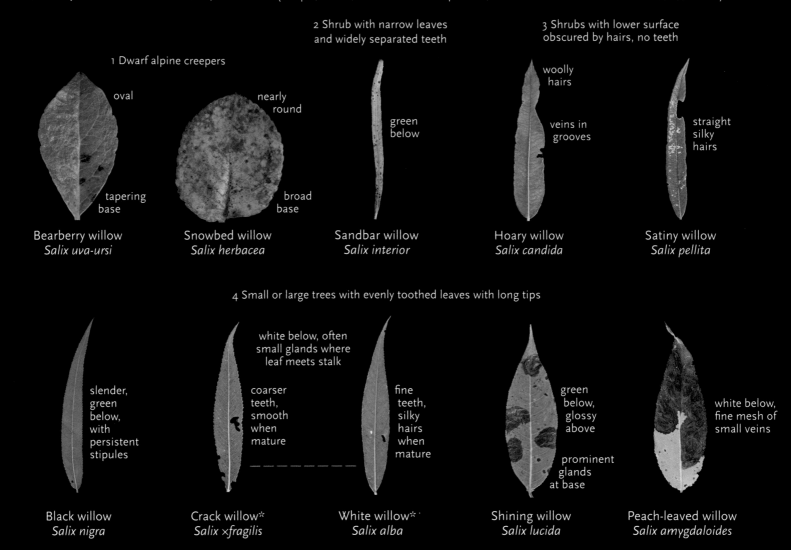

2 Shrub with narrow leaves and widely separated teeth

3 Shrubs with lower surface obscured by hairs, no teeth

1 Dwarf alpine creepers

oval

nearly round

green below

woolly hairs

straight silky hairs

tapering base

broad base

veins in grooves

Bearberry willow
Salix uva-ursi

Snowbed willow
Salix herbacea

Sandbar willow
Salix interior

Hoary willow
Salix candida

Satiny willow
Salix pellita

4 Small or large trees with evenly toothed leaves with long tips

slender, green below, with persistent stipules

white below, often small glands where leaf meets stalk

coarser teeth, smooth when mature

fine teeth, silky hairs when mature

green below, glossy above

prominent glands at base

white below, fine mesh of small veins

Black willow
Salix nigra

Crack willow*
Salix ×fragilis

White willow*
Salix alba

Shining willow
Salix lucida

Peach-leaved willow
Salix amygdaloides

THE WILLOWS have about 24 northern forest species, including several rarities not shown here. All the species are reasonably distinct, though often variable, and many have useful ecological preferences. They are best identified from mature leaves and fruits. Female flowers work pretty well too. Winter buds, staminate flowers, and immature leaves are interesting but treacherous.

Willow leaves are available, either on the plants or under them, for much of the year. When combined with ecological preferences they are diagnostic much of the time. The most useful features for the mature leaves of each species in the order they are shown above, are:

BEARBERRY WILLOW: alpine and subalpine zones, usually in protected places; small pointed leaves with tapering bases.

SNOWBED WILLOW: alpine tundra, usually in protected places; creeping shrub with small, delicate, rounded leaves with blunt tips and broad bases.

SANDBAR WILLOW: colonial shrub of river bars, also on dunes; narrow, parallel-sided leaves with sharp, widely separated teeth.

HOARY WILLOW: shrub of open limy wetlands, cedar swamps; leaves without regular teeth, white woolly hairy below, veins in grooves above.

SATINY WILLOW: uncommon tall shrub of river shores; leaves without regular teeth, satiny-white with straight hairs below.

BLACK WILLOW: abundant small or medium-sized tree with platey bark, often many-trunked, of river and lake shores and wet open ground; leaves slender, long-pointed, evenly toothed, green (rather than whitened) below, often with large persistent stipules.

CRACK AND WHITE WILLOWS: large trees, often many trunked, with braiding ridges rather than plates; leaves slender, long tipped, evenly toothed, whitened below and often with glands where the petiole meets the blade on mature leaves. White willow leaves are said to be more persistently silky than crack willow leaves, and to have finer teeth. Crack willow is a series of hybrids with the white willow as one parent; intermediates are common.

SHINING WILLOW: frequent shrub or small tree of riverbanks, shores, open and shrubby wetlands; leaves oval, long-tipped, glossy above, green below, with glands where the blade meets the petiole.

PEACH-LEAVED WILLOW: common large tree of wetlands and shores in the midwest, rare eastward; ridged bark; leaves oval, long pointed, glossy, whitened below, with a very fine meshwork of veins above.

QUICK GUIDES TO BUDS & TWIGS

QUICK GUIDES TO LEAVES

EVERGREENS

OPPOSITE BUDS

ALTERNATE BUDS

OPPOSITE LEAVES

ALTERNATE

SALICACEAE, WILLOW FAMILY ——

5 Shrubs with evenly toothed leaves, usually whitened below, that have fairly short tips

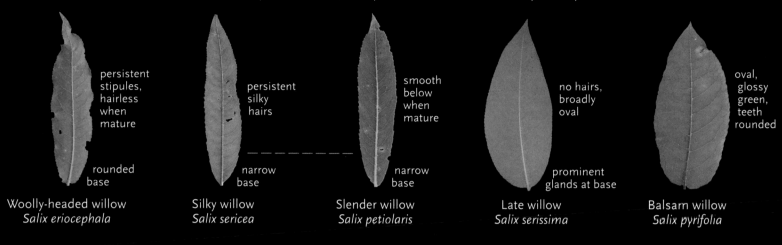

persistent stipules, hairless when mature

rounded base

Woolly-headed willow
Salix eriocephala

persistent silky hairs

narrow base

Silky willow
Salix sericea

smooth below when mature

narrow base

Slender willow
Salix petiolaris

no hairs, broadly oval

prominent glands at base

Late willow
Salix serissima

oval, glossy green, teeth rounded

Balsam willow
Salix pyrifolia

6 Shrubs with smooth leaves, whitened below, without teeth or scallops   7 Shrubs or small trees with leaves whitened below and either fur or scallops

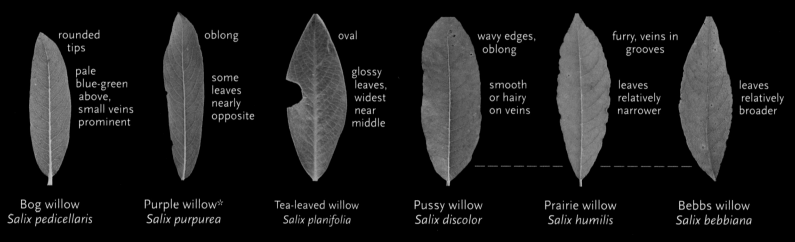

rounded tips

pale blue-green above, small veins prominent

Bog willow
Salix pedicellaris

oblong

some leaves nearly opposite

Purple willow＊
Salix purpurea

oval

glossy leaves, widest near middle

Tea-leaved willow
Salix planifolia

wavy edges, oblong

smooth or hairy on veins

Pussy willow
Salix discolor

furry, veins in grooves

leaves relatively narrower

Prairie willow
Salix humilis

leaves relatively broader

Bebbs willow
Salix bebbiana

THE SHRUBBY WILLOWS with leaves that are which below include some of our commonest species.

WOOLLY-HEADED WILLOW, also (misleadingly) called heart-leaved willow: abundant on river shores and bars, in shrub swamps, and along roads and on wet post-agricultural land; lanceolate to narrow-oval leaves, evenly toothed, usually whitened and hairless below; often with large stipules.

SILKY WILLOW: similar habitats and leaves, but silky below with straight hairs and the leaves narrower and more tapered to base.

SLENDER WILLOW: closely related to silky and in similar habitats. Sometimes distinguished by less silky leaves or red hairs on new growth; best distinguished by longer and more pointed capsules.

LATE WILLOW: Uncommon species of limy fens and cedar swamps; leaves broadly oval, finely toothed, resembling those of shining willow but whitened below and with a shorter point.

BALSAM WILLOW: occasional in northern wetlands and along roads and in clearings in boreal forest; leaves broad oval and short-pointed, glossy green above, with low teeth, on glossy bright red twigs.

BOG WILLOW: occasional in fens; untoothed leaves with blunt tips, pale green above, whitened below, veins prominent.

TEA-LEAVED WILLOW: northern species of open peaty wetlands, entering the northern forest in alpine zones in the east and wetlands in the upper midwest; leaves oval, glossy above, with wavy edges but no regular teeth; branches often red-brown.

PURPLE-WILLOW: colonial European species planted along rivers to stabilize soil; leaves narrow, untoothed except near the tip; often nearly opposite.

PUSSY WILLOW: common species of open wetlands and wet postagricultural fields; leaves large, oblong, wavy edged but not toothed, smooth or hairy, usually whitened below; twigs dark, usually furry.

PRAIRIE WILLOW: frequent in sandy barrens, prairies remnants, and dry rocky soil; leaves like pussy willow. Not always separable; generally more uniformly furry below and with the veins in grooves.

BEBBS WILLOW: common on roadsides, in disturbed and postagricultural habitats and in open wetlands. Leaves like prairie willow but somewhat shorter relative to their width, with the veins in grooves above and pronounced below.

oval leaves,
lobed or not

fringe of
tiny hairs

V

Bittersweet nightshade*
Solanum dulcamara

blunt,
light green
oboval

leafstalks
cover
buds

S

Leatherwood
Dirca palustris

ULMACEAE, ELM FAMILY: *ULMUS*, ELMS (Trees with asymmetrical, coarsely toothed leaves that are often rough or double-toothed)

some upper
veins fork

often
smooth
above

rough or
smooth above

T

American elm
Ulmus americana

very rough
above

T

Slippery elm
Ulmus rubra

corky ridges
on older twigs

T

Cork elm
Ulmus thomasii

VITACEAE, GRAPE FAMILY (High-climbing vines with tendrils) ————————————————————

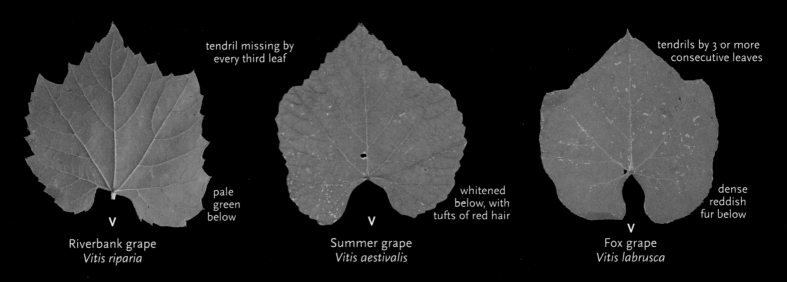

tendril missing by
every third leaf

tendrils by 3 or more
consecutive leaves

pale
green
below

V

Riverbank grape
Vitis riparia

whitened
below, with
tufts of red hair

V

Summer grape
Vitis aestivalis

dense
reddish
fur below

V

Fox grape
Vitis labrusca

QUICK GUIDES TO BUDS & TWIGS

QUICK GUIDES TO LEAVES

EVERGREENS

OPPOSITE BUDS

ALTERNATE BUDS

OPPOSITE LEAVES

ALTERNATE LEAVES

VITACEAE, GRAPE FAMILY

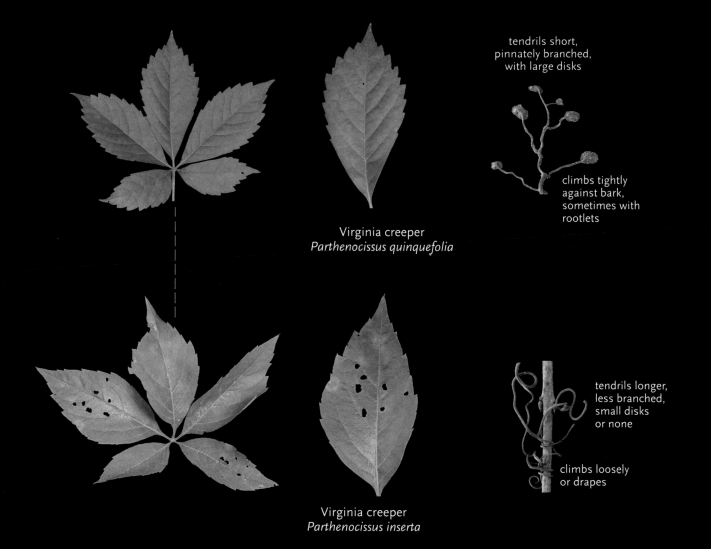

tendrils short, pinnately branched, with large disks

climbs tightly against bark, sometimes with rootlets

Virginia creeper
Parthenocissus quinquefolia

tendrils longer, less branched, small disks or none

climbs loosely or drapes

Virginia creeper
Parthenocissus inserta

THE ELMS all have similar leaves: oval, coarsely toothed (and somewhat double-toothed), and asymmetrical. They differ in how hairy, and hence how rough, they are. Slippery elm is the roughest, cork elm the smoothest. American elm, unfortunately, overlaps with both of them. The best vegetative characters seem to be the white layers in the bark of American elm, the shining red hairs on the buds of slippery elm, and the corky wings on the older branches of cork elm. The very rough leaves and symmetrically forking veins of slippery elm leaves are fairly reliable characters, but may be found in some American elms as well.

The flowers and fruits of the three are quite different: American elm has long-stalked flowers and smooth fruits with a fringe of white hairs. Slippery elm flowers are short-stalked and surrounded by red-hairy scales, and its fruits are hairy in the middle but not on the edges. Cork elm has flowers and fruits in a cluster with a central axis; the fruits are both hairy over the center and fringed at the edges.

THE GRAPES vary greatly in their overall shape and in the mount of lobing. They are separated by the color and furriness of the lower surface. Riverbank grape, a wide-ranging species, is smooth and pale green. Summer grape, a more southern species, has some cobbwebby hairs and is whitened. Fox grape, an eastern species that is also widely cultivated, has a dense, velvety blanket of short reddish hairs.

THE VIRGINIA CREEPERS are said to differ in the shininess of their leaves, the axis of the flower cluster, and the type of tendrils they commonly produce. *Parthenocissus quinquefolia,* widespread in eastern North America, is said to have duller leaves, short tendrils with adhesive disks, and a central axis to the flower cluster. *P. inserta,* a northern and western species, is said to have shiny leaves, longer tendrils without disks, and a more freely forking clusters.

These differences certainly work for some plants in some places. But because we have many sterile plants, and because none of the characters are clear-cut, they are hard to use in the field. As near as I can tell, I have both *quinquefolia* and *inserta* in my yard in eastern New York. But neither I, nor any other botanist who has visited, is certain where one ends and the other begins.

SUGAR MAPLE

SILVER MAPLE

BUTTERNUT

BUR OAK

AMERICAN ELM

BIG-TOOTHED ASPEN

SHAGBARK HICKORY

SCARLET OAK

PIN OAK

RED SPRUCE

WHITE SPRUCE

BLACK SPRUCE

WHITE PINE

PITCH PINE

JACK PINE

RED PINE

TAMARACK

WHITE PINE

INDEX